HIGH-RESOLUTION IF-TO-BASEBAND

ΣΔ ADC FOR CAR RADIOS

implement a quadrature IF-to-baseband ADC. It is shown in Chapter 4 how mismatches among the complex integrators' building blocks translate into the leakage of quantization noise power from the image band to the signal band, and vice-versa. Finally, a comparison among these architectures is presented.

Chapter 5 describes the system-level design and the circuit implementation of a single-bit 5th-order quadrature CT IF-to-baseband ΣΔ ADC for AM/FM/IBOC receivers. This 118dB DR ADC enables the realization of a car radio that does not require an IF VGA, neither an AM channel selection filter. Because of the multi-channel AM digitization, most of the AM channel selection can be performed in the digital domain.

Finally, Chapter 6 presents the conclusions of this work and a discussion about the economic feasibility of the proposed radio receiver architecture. This book ends with two appendixes. In appendix A, an analysis of the harmonic distortion in CT integrators employing MOSCAPs is presented. The input-referred noise of CT ΣΔ modulators with switched-capacitor feedback is calculated in Appendix B.

Paulo G. R. Silva
Johan H. Huijsing

Delft, October 2007.

Introduction 1

The research field investigated in this book is the design of analog-to-digital converters (ADCs) for car radio receivers. Section 1.1 first describes the market for embedded electronics in automotive products as compared to other markets. The motivation of this work is the potential cost reduction that can be achieved in car radio fabrication due to an increase on the ADC performance. Section 1.2 presents a summary of the history of the audio broadcast and the technological evolution of the car radio.

1.1 Motivation

Decades before the widespread use of electronic fuel injection, engine control and anti-lock braking systems (ABS), radios were already the primary type of embedded electronics in automotive products [1]. Most motor vehicles produced nowadays are equipped with a radio/CD player unit. Figure 1-1 shows the world annual production of automobiles [2] and personal computers (PCs) [3] from 1997 until 2005. It is estimated that the sales of car radios in the coming years will be above 60 million units per year. The car radio market is therefore equivalent to a third of the PC market. This ranks embedded automotive applications as one of the most important markets for electronic products, after cell phones, PCs and home entertainment.

The car radio market is characterised by high volume production and low profit margins. In this type of market, the best way to further reduce fabrication cost is by decreasing the number of components and thus reducing the complexity of the entire system. An increase of the level of integration of the radio receiver is therefore a fundamental step towards reducing fabrication cost.

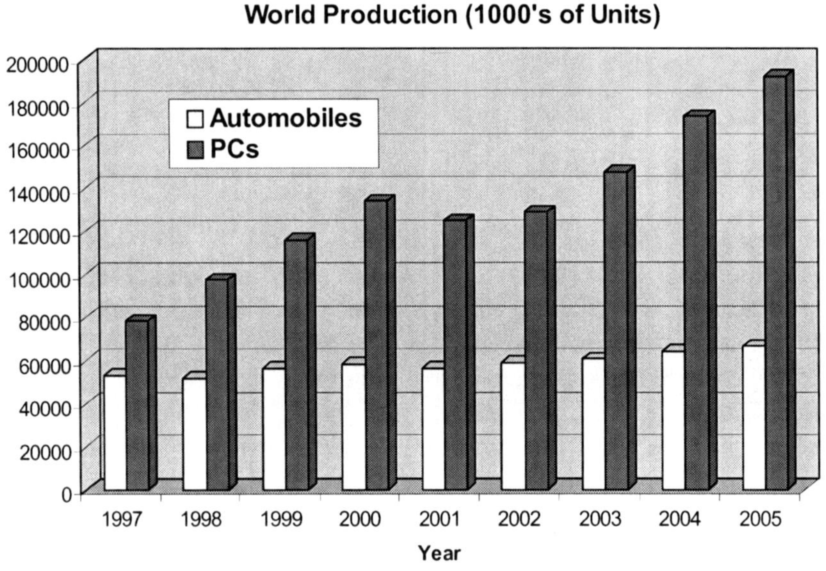

Figure 1-1: *World production of automobiles and PCs [2]–[3].*

Figure 1-2 shows the architecture of a typical DSP-based AM/FM radio receiver [4–5]. Two separate radio front-ends mix AM and FM signals to a 10.7MHz intermediate frequency (IF). The IF signals are filtered by ceramic AM and FM channel selection filters and amplified by a high-linearity/low-noise variable gain amplifier (VGA). The VGA provides the analog input for a 100dB dynamic range (DR)

IF-to-baseband ADC. The IF-to-baseband ADC then down-converts the analog input to a low-IF and provides the DSP with in-phase (I) and quadrature-phase (Q) digital inputs. The AM and FM front-ends, tuning circuitry and the RF mixers are integrated in a single BiCMOS tuner IC, while the IF-to-baseband ADC and the DSP are integrated in another digital CMOS IC.

The most expensive non-integrated components in this receiver are the AM and FM ceramic filters [6], and the VGA. Table 1-1 shows some characteristics of those components. The AM filter is the most expensive component while the VGA power consumption [7] is approximately 10 times greater than the consumption of the whole IF-to-baseband ADC [5].

The use of external filters and a VGA relaxes the performance requirements on the ADC (Chapter 2). In order to simplify the radio system and reduce the number of external components, an ADC with higher performance is required. For instance, if the costly AM filter and VGA are to be removed, the VGA's dynamic range has to be incorporated into the ADC's dynamic range. Figure 1-3 shows the proposed AM/FM receiver with a 118dB DR ADC and just one channel selection filter [8]. Based on the annual sales of car radios and on Table 1-1, it is estimated that such a radio receiver architecture would enable an economy of about US$27 million per year with external components. Further cost reduction can be achieved by simplifications on the component logistic/handling, reduction of the PCB area, etc.

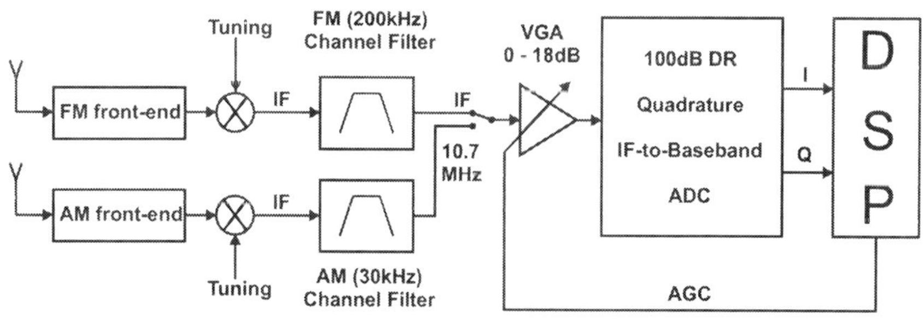

Figure 1-2: *State-of-the-art DSP based AM/FM receiver for car radios [4]–[5].*

Table 1-1. *Characteristics of some car radio non-integrated components.*

Component	Center freq.	BW	Power	Unity price
AM filter	10.7 MHz	30 kHz	–	0.35 US$
FM filter	10.7 MHz	200 kHz	–	0.10 US$
VGA	10.7 MHz	>200 kH	230 mW	0.10 US$

The improvement of the IF-to-baseband ADC's performance also allows the use of more digital signal processing in the receiver. For instance, the 118dB DR IF ADC in Figure 1-3 digitizes about 20 AM channels at the same time. This means that in AM mode, most of the channel selection is performed in the digital domain. The more channel filtering is performed in the digital domain, the more flexible a receiver becomes. The ultimate DSP based receiver would rely exclusively on digital channel filtering and could be reconfigured by software to work with different modulation schemes and/or as a part of different communication systems [9]. The proposed receiver architecture is therefore an intermediate step towards the implementation of the software-defined radio concept.

The analysis, the system-level design and the circuit implementation of the 118dB DR IF-to-baseband ADC [8] that enables the realization of the AM/FM radio receiver architecture shown in Figure 1-3 are the main subjects of this book.

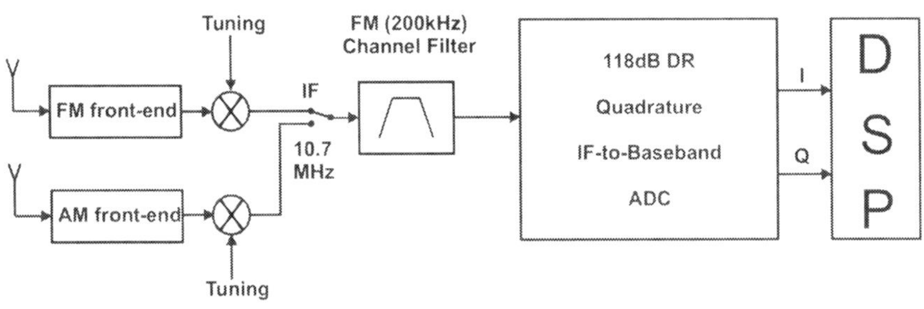

Figure 1-3: *Proposed DSP based AM/FM receiver for car radios with reduced number components [8].*

1.2 Car Radio History

The history of commercial radio broadcasting started in 1920 with the first licensed AM station in North America. As early as 1922, Daimler explored the feasibility of an AM vacuum tube receiver installed in the back compartment of an automobile. The first practical commercial car radio was just introduced in 1929 by Galvin. He named it "Motorola" (from MOTOr VictROLA). However, at this early stage of car radio technology, it was not possible to listen the radio while driving due to several electro-mechanical compatibility problems and sources of interference. Great improvements on the receiver performance occurred in the 1930s. These include the adoption of superheterodyne architecture, automatic volume control, radios that could fit on the car dashboard and the use of steel rod antennas. Before the beginning of World War II, the car radio had evolved into a reliable and affordable product. About 20% of all manufactured cars had factory installed radios [1]. An example is the Philco model 802 car radio (Figure 1-4), that was commercialised during the 1940s [10].

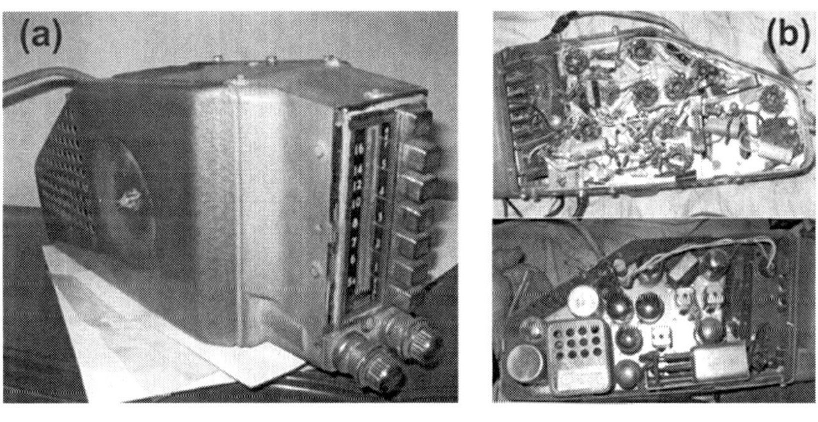

Figure 1-4: *Philco model 802 car radio (a) and its internal components (b). (Courtesy D. Froehlich [10])*

The next big improvement in the quality of the radio reception occurred as a result of the introduction of the FM by Armstrong. The first commercial FM station started broadcasting around 1941 in North America, within the 42–50 MHz band. After the World War II FM broadcasting was moved to the 87.5 to 108.5 MHz band. In 1948, another revolution took place with the invention of the transistor at the Bell Labs. The transistorization of the receiver allowed significant miniaturization and cost reduction. By 1957, the first completely transistorized FM car radio was commercialised.

During the 1960s and 1970s, the trend towards miniaturization followed the evolution of solid-state electronics technology. In 1963, the first AM/FM car radio was introduced, and in 1969, the first AM/FM-stereo radio with Cassette player was commercialised. An example of miniaturization is the super micro radio-wristwatch shown in Figure 1-5. This prototype was developed by Sony in 1982, based on a bipolar technology single-chip analog AM/FM radio receiver [11].

By the mid-1980s, due to the availability of the first cheap digital signal processors, several new features were added to the car radio. Already in 1987, CD players were integrated with the car radio as a single factory installed device. At this last stage of the car radio evolution, new digital features were mainly responsible for the improvement of the radio user interface and quality of reception, as the old analog modulation techniques remain unchanged. Table 1-2 summarizes the key events in the history of car radio.

Figure 1-5: *Super micro radio-wristwatch prototype, developed by Sony in 1982.* (*Courtesy IEEE [11]*)

Table 1-2. *Key events in the history of car radio [1].*

Year	Event
1920	First AM broadcast station
1922	Daimler demonstrated car radio
1926	First portable vacuum tube AM radio
1927	Able to listen to radio with the motor running
1929	First practical car radio
1930	Supeheterodyne introduced
1931	Dynamic loudspeaker; automatic volume control
1933	Ford introduced tailor-made radio to fit car dashboard
1937	Push button tuning; steel rod antenna
1938	Telescopic antenna
1941	First FM broadcast station
1947	First successful signal seeking radio
1948	Invention of the transistor
1956	First car radio using high-power transistors
1957	First completely transistorized FM car radio
1961	FM stereo broadcasting introduced
1963	First AM/FM car radio
1966	8-track stereo tape player introduced
1969	AM/FM-stereo/Cassette in one package
1977	AM/FM-stereo/CB transceiver
1984	Portable and car CD player introduced
1987	Factory radio with CD playback
1995	First DAB broadcasting in Europe
1996	Introduction of DVD technology
1997	IBOC standard is defined
2001	XM radio roll-out in the U.S.
2002	Sirius satellite radio roll-out in the U.S.
2002	IBOC broadcasting is approved by FCC
2006	First factory installed IBOC compatible car radio

Figure 1-6: *XM satellite car radio receiver from Delphi.*

The most recent advances in car radio technology are related to the introduction of digital modulation schemes for audio terrestrial broadcasting. The first attempt at full digital broadcasting was the digital audio broadcasting (DAB) Eureka-147 standard. The DAB service is broadcast on the 174–240MHz band and is incompatible with standard AM/FM radios. The most important features of the DAB standard are transmission of digitally encoded high-fidelity audio and immunity to multi-path interference. DAB broadcasting in Europe started in 1995, but up to now it has had a much smaller audience than FM radio. Digital audio broadcasting was also developed for satellite transmission. In 1997 the SDARS (Satellite Digital Audio Radio Service) was licensed by the FCC to the operators XM radio and Sirius. They transmit in the 2.3-GHz S band, from 2,320 to 2,345 MHz. However, SDARS broadcasting is not a free service. Both XM radio and Sirius require user subscription and payment of monthly fee. Figure 1-6 shows an example of commercial XM car radio receiver.

In the USA the DAB standard for terrestrial broadcasting was not accepted by the FCC. The main reason for this was the lack of backwards compatibility with the traditional AM/FM services. Instead, the IBOC (In-Band, On-Channel) standard was developed during the 1990s and approved by 2002. IBOC provides a digital broadcasting service that is fully compatible with the old radios. The digital information is transmitted as sidebands placed adjacent to the traditional AM or FM channels. IBOC is described in more detail in Chapter 2.

As more computing power is becoming available inside the automobile, the car radio is evolving into a multimedia processing unit. On-line (terrestrial broadcasting, satellite radio, GPS navigation, internet) and off-line (CD, DVD, USB storage) content will be available individually to each passenger in the vehicle. Multi-channel receivers together with hard drive storage will enable tuning of different stations by each passenger, as well as the recording of real-time channels for later listening as personalised off-line content.

References

1. R.C. Lind, H.W. Yen and D.L. Welk, "Evolution of the Car Radio - From Vacuum Tubes to Satellite and Beyond" *SAE Internat. Congress & Expo. on Transp. Electronics*, Oct. 2004.
2. World Automotive Industry Production Statistics, available on-line at www.oica.net
3. Personal Computer Production Statistics: 1975–2005, available on-line at www.pegasus3d.com/total_share.html
4. E.J. van der Zwan, K. Philips and C.A.A. Bastiaansen, "A 10.7-MHz IF-to-Baseband Sigma Delta A/D Conversion System for AM/FM Radio Receivers", *IEEE J. Solid-State Circ.*, vol. 35, 1810–1819, Dec. 2000.
5. Q. Sandifort, L.J. Breems, C. Dijkmans and H. Schuurmans, "IF-to-Digital Converter for FM/AM/IBOC Radio", Proceedings of the 29th ESSCIRC, Sept. 2003, Estoril, Portugal, pp. 707–710.
6. CERAFIL catalogue, available on-line at www.murata.com
7. Catena Radio Design, private communication, 2005.
8. P. Silva, L. Breems, K. Makinwa, R. Roovers and J. Huijsing, "An IF-to-Baseband ΣΔ Modulator for AM/FM/IBOC Radio Receivers with a 118 dB Dynamic Range", *IEEE J. Solid-State Circ.*, vol. 42, 1076–1089, May 2007.
9. E. Buracchini, "The Software Radio Concept", *IEEE Commun. Mag.*, Sept. 2000.
10. Philco Model 802, Antique Radio and TV Restoration Home Page, www.tubesandtransistorsandmore.com.
11. T. Okanobu, T. Tsuchiya, K. Abe and Y. Ueki, "A Complete Single Chip AM/FM Radio Integrated Circuit", *IEEE Trans. Consumer Electron.*, vol. 28, 393–407, Aug. 1982.

2

DSP Based Radio Receiver Architectures

The incorporation of digital signal processing into the receiver has made the implementation of fully digital demodulation schemes possible. Such schemes are an effective way to improve the reception quality in legacy analog broadcasting systems.

In a scenario where the analog signal processing is to be replaced by digital functions, the ADC is a key part of a modern receiver. Several DSP based receiver architectures are described in section 2.1. The ADC performance, i.e. dynamic range, resolution, linearity, etc. determines the location of the ADC in the receiver path. These performance metrics are defined in section 2.2 and the concepts of dynamic-range extension, desensitization and blocking are discussed in section 2.3. Section 2.4 reviews the issue of image rejection in low-IF and zero-IF receivers, with emphasis on the improvements achieved by quadrature architectures. Section 2.5 presents the technical specifications of analog radio broadcasting and several examples of integrated architectures for AM/FM receivers. The technical specifications and the most important features of the IBOC digital radio broadcasting standard are presented in section 2.6. This chapter ends in section 2.7 with a description of several commercial chip-sets for AM/FM and IBOC radios.

2.1 Radio Receiver Architectures

From the early days of the AM radio broadcasting to the widespread use of mobile phones, the evolution of the wireless systems has followed the development of electronic technology. While both transmitters and receivers have profited from this continuous development, the latter have undergone the most revolutionary changes due to demands for increased portability and reduced power consumption. The replacement of vacuum tubes by transistors after World War II was responsible for the first radical changes leading to cost reduction and miniaturization. Further miniaturization was achieved with increased integration and digitization of the receiver back-end. However, even today, external passive components are necessary in the radio front-end. In a modern receiver, the amounts of analog and digital signal processing, and the level of integration are dictated by the characteristics of different communication systems. Each case requires the hardware to be developed with a different emphasis on cost reduction, performance (bandwidth, dynamic range, linearity, etc.), portability and programmability.

2.1.1 Heterodyne and Homodyne Receivers

Figure 2-1 portrays an example of the RF input spectrum received by the antenna. This spectrum is composed of the entire receiver band plus some interferers. The purpose of the receiver is to retrieve the information within a desired channel with a minimum pre-defined quality level. In order to accomplish this task, several steps of amplification, filtering and down-conversion may be used [1].

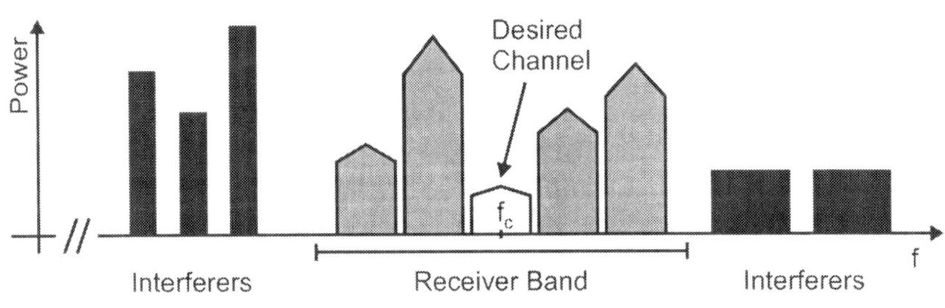

Figure 2-1: *Receiver RF input spectrum.*

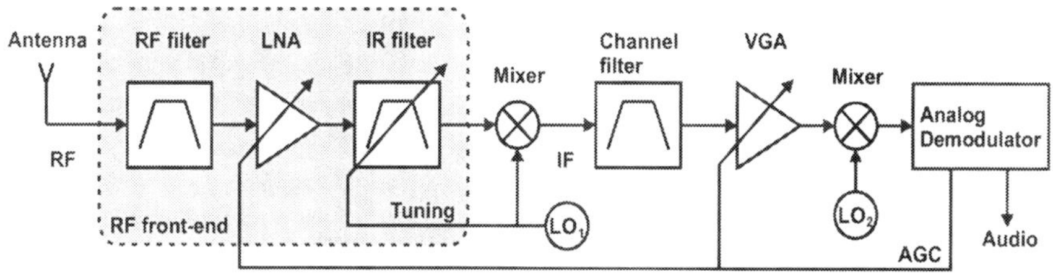

Figure 2-2: *Fully analog heterodyne receiver [1].*

Figure 2-2 shows an example of a fully analog heterodyne receiver. Because the receiver band may be surrounded by strong interferers, a first RF bandpass filtering step is required after the antenna. The received signal may be very weak and needs to be amplified by the low noise amplifier (LNA) prior to further signal conditioning. To filter the desired channel directly from the RF, very high quality-factor filters would be needed. In order to relax the quality-factor requirements for channel selection, a down-conversion mixer is used to translate the receiver band to a lower intermediate frequency (IF). The tuning of the receiver to select the desired channel is performed by changing the frequency of the first local oscillator (LO_1). The mixer stage however, requires an image rejection filter to attenuate interferers located at the image frequencies. At IF frequencies, the desired channel can be selected and all other channels are strongly attenuated by the channel filter. The desired channel is then amplified again by the variable gain amplifier (VGA) and then down-converted to a lower frequency. The final processing step is analog demodulation, where the baseband analog information is retrieved. In order to increase the receiver's dynamic range, both the LNA and VGA gains are controlled by an automatic gain control (AGC) loop [1].

When the LO_1 is chosen to be equal to the center frequency (f_C) of the desired channel, this channel is directly down-converted to dc and no other mixer stages are needed. This receiver architecture is known as homodyne or zero-IF, and is depicted in Figure 2-3. Because the IF frequency is dc, the image band contains a replica of the desired channel. The homodyne architecture is sensitive to low-frequency errors, such as dc offset (from amplifiers or caused by self-mixing) and *1/f* noise, which fall inside the desired channel [1]. The issue of zero-IF mixing and image rejection is discussed in more detail in section 2.4.

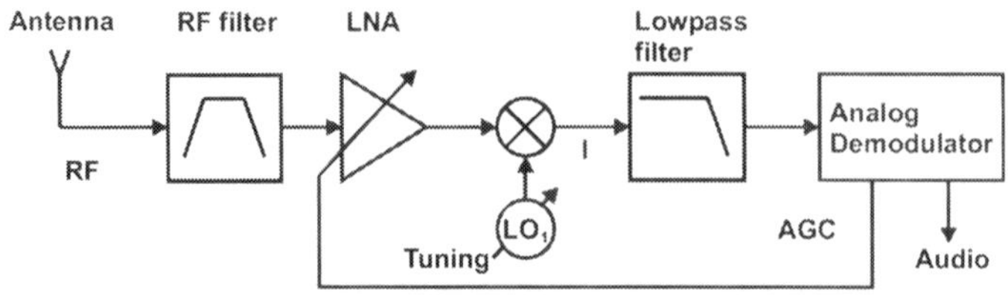

Figure 2-3: *Fully analog homodyne receiver [1].*

2.1.2 DSP Based Radio Receiver Architectures

The continuous development of CMOS technology and the increasing availability of digital processing power has enabled a series of mixed analog/digital radio receivers. With the first generation of DSP based receivers, the only analog operations to be implemented in software were audio baseband functions such as stereo decoding and audio control. These digital features can be added to the receiver architectures shown in Figures 2-2 and 2-3 if an ADC and a DSP [8]–[10] replace the analog audio processing circuitry. Several high quality-factor filters are typically still necessary for the full analog channel selection [11], which limits the level of integration.

The next generation of DSP based receivers (Figure 2-4) incorporated digital demodulation to replace analog demodulation techniques [12]. This digital processing greatly improved the quality of the reception using several new features that could not be easily implemented in fully analog radios, such as adjacent channel suppression and multi-path interference detection/suppression. The performance required from an ADC in a receiver architecture depends on the amount of analog signal processing operations that precede the digitization. In the architecture shown in Figure 2-4, several stages of filtering, amplification and down-conversion relax the ADC's speed and resolution requirements. In order to increase the level of integration and reduce costs, some of the external channel selection filters can be replaced by integrated active filters. Because of the limited quality-factor, quadrature down-conversion is normally used to relax image rejection requirements (see section 2.4) of the integrated active filters. However, two independent in-phase and quadrature-phase (I/Q) paths and two ADCs are required by this architecture.

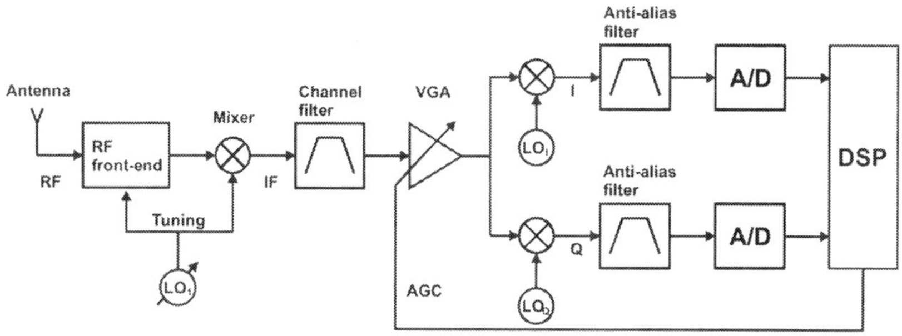

Figure 2-4: *Heterodyne quadrature receiver with baseband ADCs and digital demodulation [12].*

Another way to further reduce the number of external analog components is to move the ADC to the IF level, as shown in Figure 2-5. In this architecture, a selected narrow-band channel is digitized at IF and the quadrature mixing is performed in the digital domain with almost perfect linearity and I/Q matching [13]–[17]. Because the signal at IF is not complex, just one ADC is required. However, the direct IF digitization requires a faster ADC than the previous architecture (Figure 2-4). A high quality-factor channel filter and an AGC loop relax the DR and bandwidth requirements of the ADC. The IF ADC in this architecture is normally an oversampled bandpass ΣΔ ADC [18]. Bandpass ΣΔ ADCs are discussed in more detail in section 4.4.

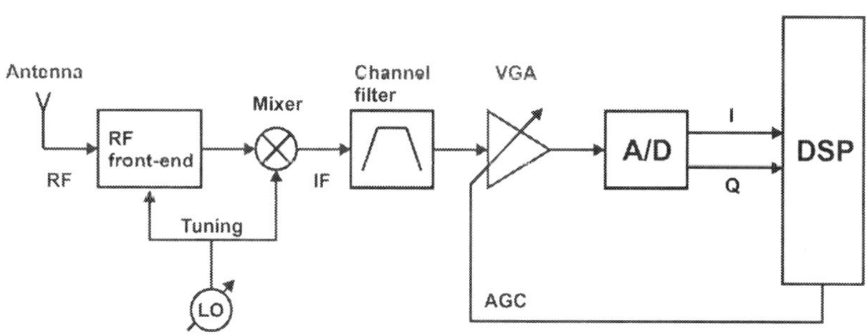

Figure 2-5: *Heterodyne receiver with bandpass IF ADC and digital demodulation [13]–[17].*

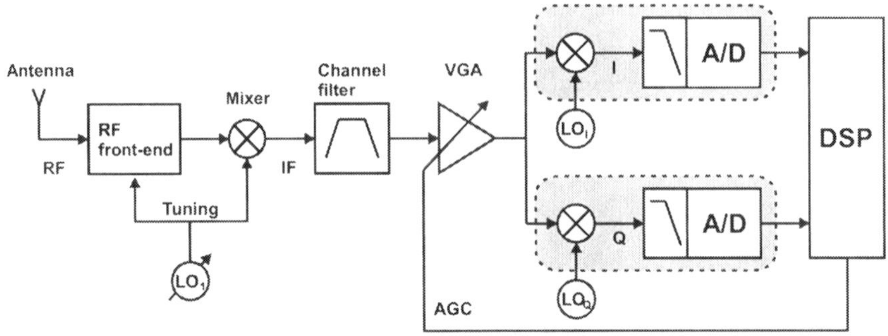

Figure 2-6: *Heterodyne receiver with IF-to-baseband ADCs and digital demodulation [19]-[21].*

An alternative approach to IF digitization was achieved by the combination of integrated down-conversion mixers and baseband ADCs. The receiver architecture shown in Figure 2-6 combines the decreased number of external filters offered by the IF digitization architecture with more linear and power efficient IF-to-baseband ADCs [19]. The drawback of this architecture is that the analog quadrature down-mixing is prone to I/Q mismatch and limits the image suppression (section 2.4). Figure 2-7 shows a receiver with higher DR IF-to-baseband ADCs [23] that do not need to be preceded by a VGA. The removal of the AGC loop prevents strong interferers from desensitizing the receiver when the desired channel is very weak (section 2.3).

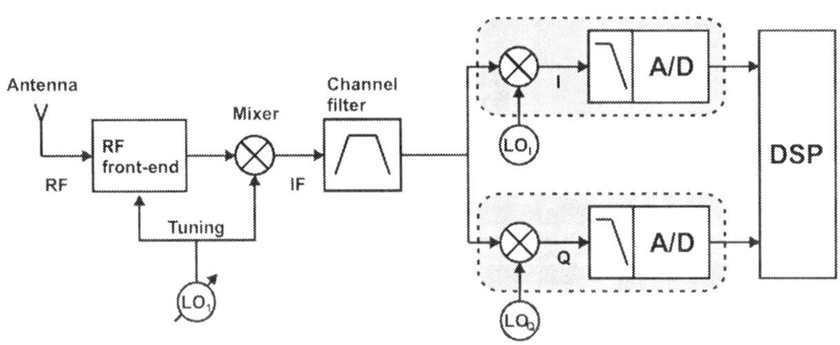

Figure 2-7: *Heterodyne receiver with higher dynamic range IF-to-baseband ADCs and digital demod. [23].*

2.2 ADC Performance Metrics

The most important performance metrics for ADCs used in telecommunication applications are reviewed in this section [24]. They are divided in two groups: resolution definitions and linearity definitions.

2.2.1 Measures of Resolution

The resolution of an ADC expresses the minimum detectable change of the analog input related to the maximum input. The most important measures of resolution are dynamic range, signal-to-noise ratio and effective number of bits. Figure 2-8 shows the signal-to-noise ratio plotted as a function of the power of a sinusoidal input, and the relation between several measures of resolution.

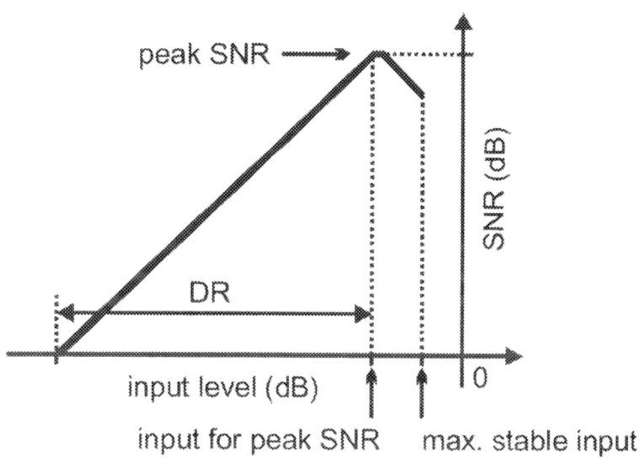

Figure 2-8: *Definition of dynamic range (DR).*

Dynamic range **(DR)** – ratio between maximum input power and minimum detectable input power within a certain bandwidth. Generally the minimum input is determined by the noise power in the bandwidth of interest.

Signal-to-noise ratio **(SNR)** – ratio between the input power (normally a sinusoidal signal) and the noise power inside a certain bandwidth.

Peak **SNR** – peak value of the SNR plot. The ADC resolution is very often expressed by the peak SNR in number of bits.

***Effective number of bits* (ENOB)** – the peak SNR at the ADC output expressed as a number of bits:

$$ENOB = \frac{SNR_{peak}(dB) - 1.76}{6.02} \tag{2-1}$$

2.2.2 Measures of Linearity

ADC linearity is a very important performance specification in DSP based receivers. Receiver architectures with multi-band digitization are especially sensitive to distortion components induced by strong unwanted channels inside the bandwidth of a weak wanted channel. Ideally, non-linearity induced spectral components should be below the minimum detectable input signal. The most important linearity definitions are harmonic distortion, intermodulation distortion, spurious-free dynamic range, signal-to-noise-and-distortion ratio (SNDR), peak SNDR, intermodulation intercept point and cross-modulation distortion.

***Harmonic distortion* (HD_x)** – ratio between maximum input sinusoidal power and the power of the xth harmonic of the input tone. The second (HD_2) and the third (HD_3) harmonic components are normally the most important (Figure 2-9).

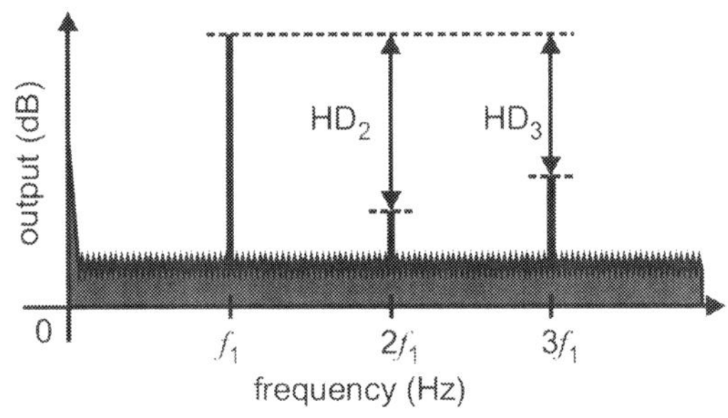

Figure 2-9: *Definition of harmonic distortion (HD_x).*

***Intermodulation distortion* (IM_x)** – defined for a maximum power two tone input test, where distortion components due to non-linearity are present in spectral positions which are combinations of the input signal frequencies f_1 and f_2. The intermodulation distortion is defined as the ratio between the power of one of the signal tones and the power of xth-order intermodulation distortion tone. The second (IM_2) and the third (IM_3) intermodulation components, respectively located at the spectral positions (f_2-f_1),(f_1+f_2) and (2f_1-f_2),(2f_2-f_1), are the most important (Figure 2-10).

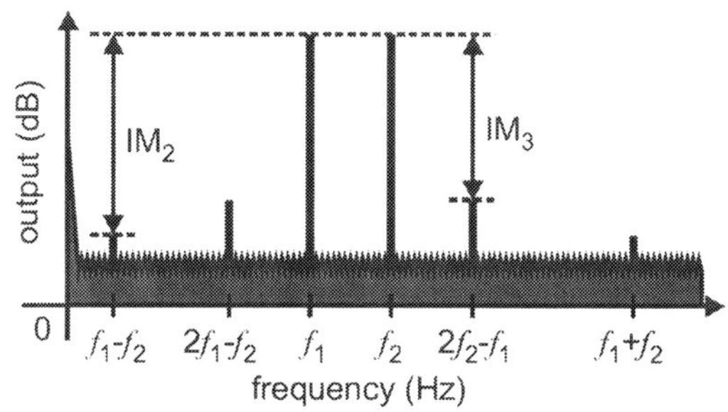

Figure 2-10: *Definition of intermodulation distortion (IM_x).*

Spurious-free dynamic range **(SFDR)** – ratio between maximum power input sinusoidal and the strongest in-band spurious tone power (Figure 2-11).

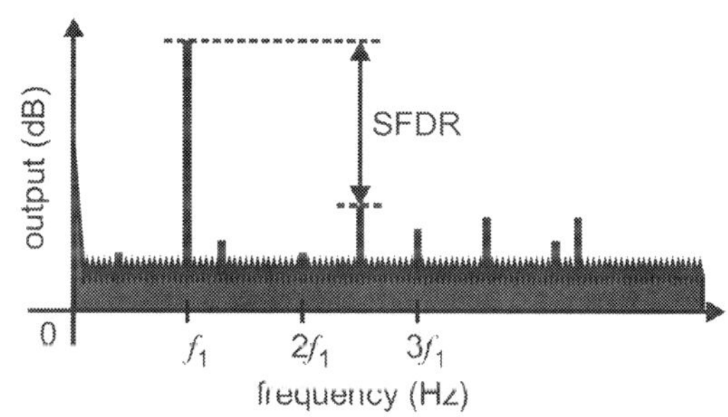

Figure 2-11: *Definition of spurious free DR (SFDR).*

Signal-to-noise-and-distortion ratio **(SNDR)** – ratio between the input power (normally a sinusoidal signal) and total noise and distortion power inside a certain bandwidth.

Peak **SNDR** – ratio between the power of the sinusoidal input for peak SNR and the total noise and distortion power inside the ADC bandwidth.

***Intermodulation intercept point* (IP$_x$)** –theoretical sinusoidal input carrier power for which the xth-order intermodulation product is equal to the signal carrier power. IP$_2$ and IP$_3$ (Figure 2-12) are the most important.

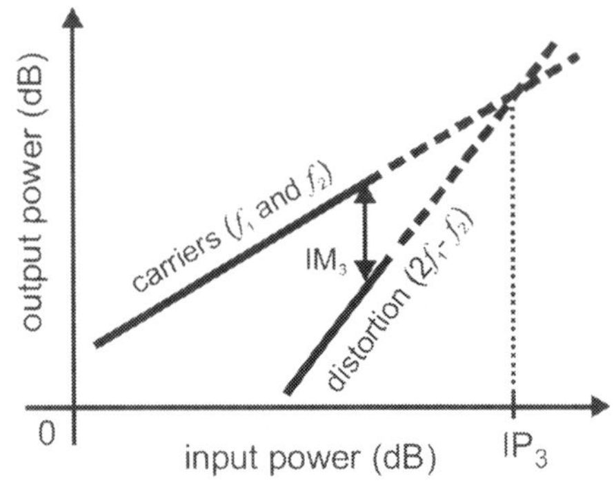

Figure 2-12: *Definition of IP$_3$.*

***Cross-modulation distortion* (CM)** – modulation of the spectrum around the carrier (f_1) of the wanted channel by the spectral content of an unwanted channel due to non-linearity (Figure 2-13). Cross-modulation distortion is defined as the distance between the desired carrier and the strongest cross-modulation distortion component.

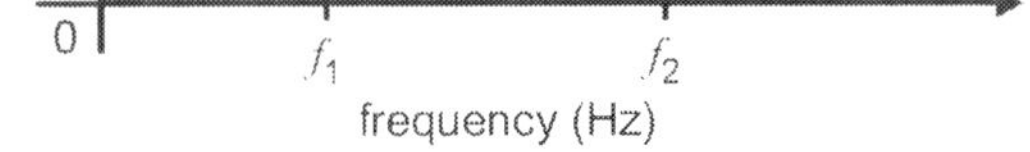

Figure 2-13: *Definition of cross-modulation distortion (CM).*

2.3 Desensitization and Blocking

The input-referred DR of an ADC can be extended if it is preceded by a variable gain amplifier and a filter (Figure 2-14).

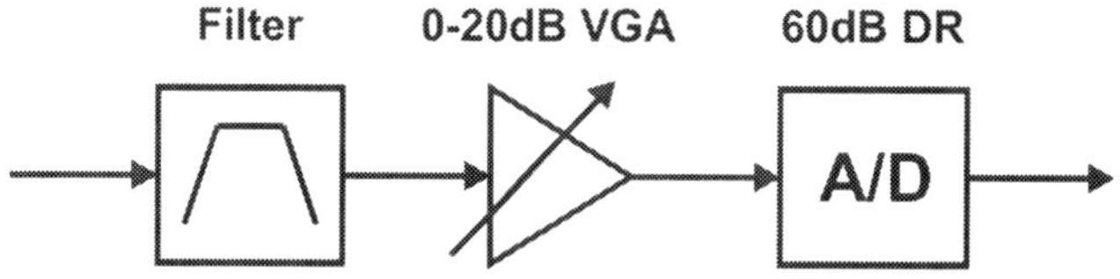

Figure 2-14: *ADC with extended dynamic range.*

The input-referred DR of a 60dB DR ADC combined with a VGA with 3 programmable gains steps (0, 10, 20 dB) is shown in Figure 2-15. In this picture the SNR for a sinusoidal input is plotted for several amplitudes, varying from 0 to -80dB_{FS}. For larger inputs, the VGA gain is set to 0dB and the peak SNR of 60dB is achieved. For very small signals, the ADC input is amplified by 20dB, extending the input-referred DR from 60 to 80 dB. The combination VGA+ADC works properly if the VGA is perfectly linear and no strong interferers are present.

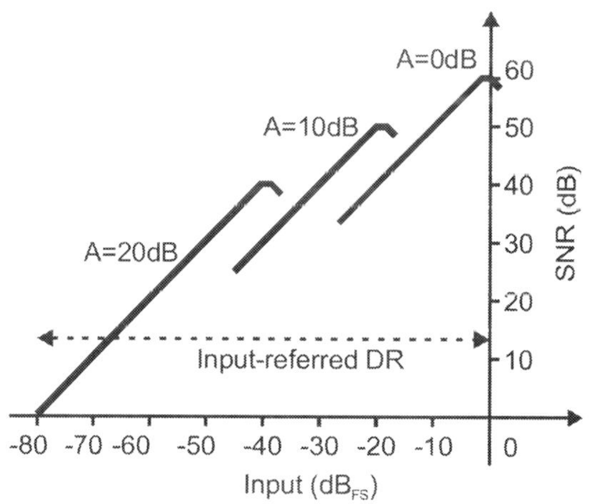

Figure 2-15: *Input DR of the combination VGA + ADC.*

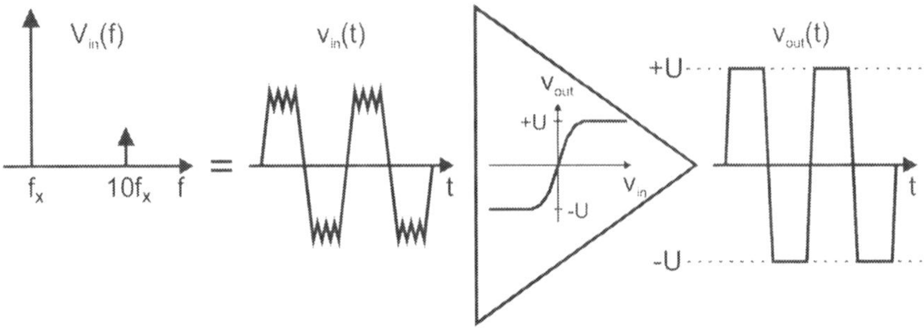

Figure 2-16: *Saturated output of an amplifier with a compressive characteristic (worst case).*

However, the situation described in the last sentence of the previous paragraph is unrealistic: strong interferers and adjacent channels are often present in the received spectrum, and real amplifiers present non-linearities like saturation. Figure 2-16 shows a typical amplifier with a (non-linear) compressive input-output characteristic with a two-tone input and a clipped output. If the desired information is carried by the weak tone at a high frequency and the strong tone is part of an interference channel, the large amplitude signal forces the amplifier to saturate. In the worst case, the high-frequency tone is not present at the output (gain 0). In this situation the receiver is said to be desensitized and the desired signal is blocked by the strong interference [1].

Therefore, in a practical receiver, the combination VGA+ADC is always preceded by filtering as in Figure 2-14. The effect of a filter preceding an AGC controlled VGA is better understood with the aid of Figure 2-17. The wanted channel centred at the IF is surrounded by unwanted adjacent channels. When the wanted channel is very strong the AGC loop sets the gain of the VGA to 0dB (Figure 2-17a). The unwanted channels are strongly attenuated by the filter and the desired channel is not blocked. Figure 2-17b shows the other extreme case: the wanted channel is so weak that the VGA is set to its maximum gain. The wanted signal can still be properly amplified because strong adjacent channels are filtered out before the VGA. In the case the filter is absent, the whole receiver could be desensitized by strong adjacent channels like the amplifier in Figure 2-16.

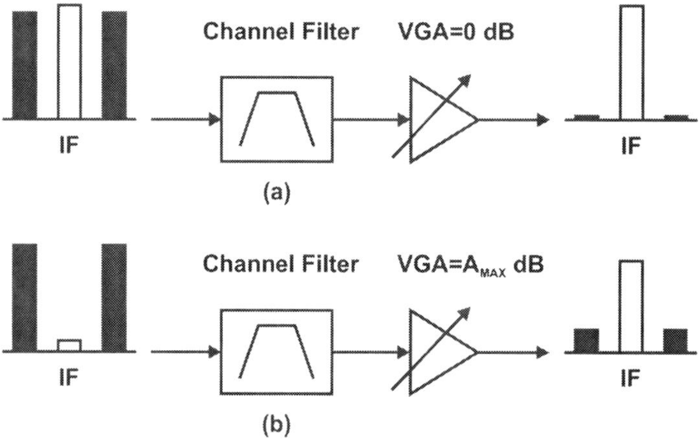

Figure 2-17: *Combination of a filter and a VGA: strong desired channel and VGA=0dB (a), weak desired channel and* $VGA=A_{MAX}$ *dB (b).*

In Figure 2-17 it is assumed that no interferers are present in the transition band of the filter. Figure 2-18 shows the common situation when some adjacent channels are present in the frequency bands where the filter attenuation is not strong. This is the case of a narrow-band filter with limited quality-factor centred around a high IF, for example an AM channel selection filter with 30 kHz bandwidth for the 10.7 MHz IF. The final channel selection is often performed in the digital domain to compensate for the lack of attenuation around the desired channel. When the wanted channel and the adjacent channels are all weak (Figure 2-18a), the maximum VGA gain is selected and the wanted channel is properly received. However, when the adjacent channels are very strong (Figure 2-18b, neglecting harmonic distortion), the insufficiently attenuated adjacent signals can saturate the VGA and significantly reduce the amplification of the wanted channel. In an extreme case, the wanted channel can be blocked. The sensitivity of the receiver can be improved by a higher quality-factor channel filter. However, this option increases the overall cost of the radio. The most elegant solution to increase the receiver sensitivity is to use an ADC with higher DR, in such a way that the VGA and AGC loop are not required anymore. Of course, this solution poses some challenging DR specifications on the ADC design (see Chapter 5).

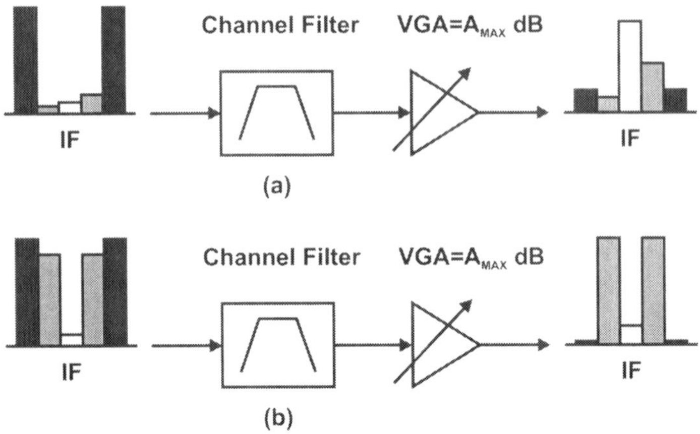

Figure 2-18: *Filter and VGA with weak desired channel surrounded by non-attenuated weak adjacent channels (a), by strong adjacent channels (b).*

2.4 Image Rejection

All architectures discussed in section 2.1 rely on one or more down-conversion stages to successfully select the wanted channel. The single-path (real) mixing operation (shown in Figure 2-19a) is described mathematically by a multiplication of the mixer RF input spectrum with a single-tone sinusoidal at the radian frequency ω_{LO} [1]:

$$RF \times LO = IF = \cos(\omega_{RF}t) \cdot \cos(\omega_{LO}t) = \frac{\cos[(\omega_{RF}-\omega_{LO})t]}{2} + \frac{\cos[(\omega_{RF}+\omega_{LO})t]}{2} \quad (2\text{-}2)$$

Equation (2-2) reveals that the input RF spectrum is down-converted to an IF and up-converted to a higher RF. In a receiver, the up-converted components are not a problem if the mixing operation is followed by lowpass filtering. More important is the fact that the RF frequency components located at ($\omega_{LO}+\omega_{IF}$) and ($\omega_{LO}-\omega_{IF}$) are both converted to the same IF. If the first is the center frequency of the signal band, the second is named the image band, and vice-versa. Figure 2-20 shows the image problem in single-path low-IF (Figure 2-2) and zero-IF (Figure 2-3) receivers.

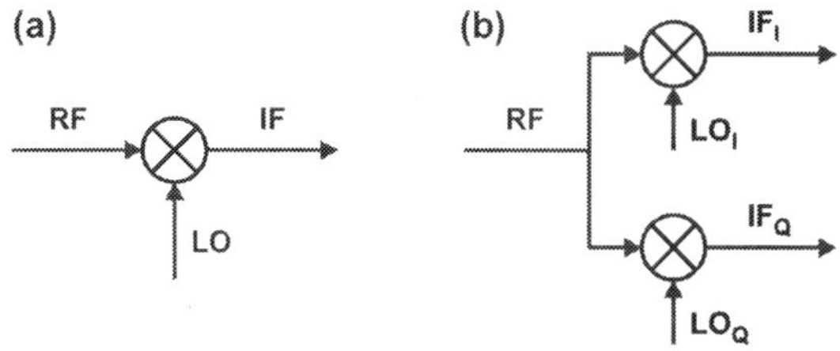

Figure 2-19: *Single-path (real) mixing (a), quadrature-paths (complex) mixing (b).*

Figure 2-20a describes the image problem in the low-IF down-conversion: if the power at the image band is stronger than the signal channel, the wanted information is completely corrupted. This is the reason why single-path mixers have to be preceded by a high selective and often tuneable image rejection filter in heterodyne receivers (Figure 2-2). In the case of single-path zero-IF down-mixing (Figure 2-20b), the image band contains a mirrored version of the wanted band. Both are converted to dc and the desired information is always corrupted if a single sideband (SSB) modulation scheme, as in citizens band (CB) amateur radio, is not used [1].

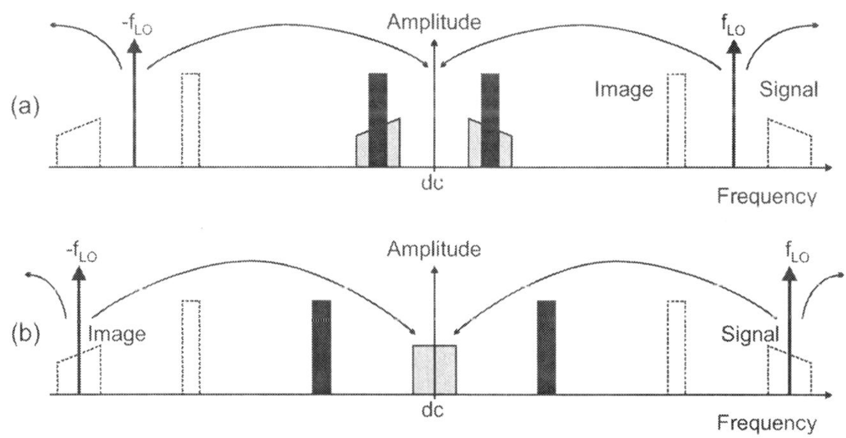

Figure 2-20: *Single-path (real) low-IF down-conversion (a) and single-path zero-IF down-conversion (b).*

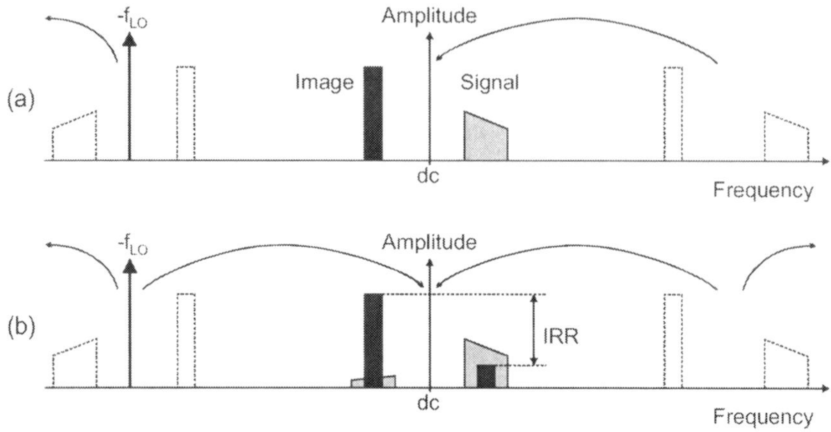

Figure 2-21: *Quadrature-paths low-IF down-conversion with ideal matching (a), with I/Q mismatch (b).*

The quadrature (complex) mixing operation depicted in Figure 2-19b is able to alleviate the image problem. The mixers' oscillator in-phase (LO_I) and quadrature-phase (LO_Q) inputs are 90° phased from each other:

$$LO_I = \cos(\omega_{LO}t) = \frac{e^{j\omega_{LO}t}}{2} + \frac{e^{-j\omega_{LO}t}}{2}$$
$$LO_Q = -\sin(\omega_{LO}t) = \frac{e^{j\omega_{LO}t}}{-2j} + \frac{e^{-j\omega_{LO}t}}{2j} \qquad (2\text{-}3)$$

The signals LO_I and LO_Q can be considered the real and imaginary parts of the complex signal $LO = LO_I + jLO_Q$. The multiplication of the real RF input with LO results in another complex signal at the IF output:

$$LO = \cos(\omega_{LO}t) - j\sin(\omega_{LO}t) = e^{-j\omega_{LO}t} \qquad (2\text{-}4)$$

$$RF \times LO = IF = \cos(\omega_{RF}t) \cdot e^{-j\omega_{LO}t} = \frac{e^{j(\omega_{RF}-\omega_{LO})t}}{2} + \frac{e^{j(-\omega_{RF}-\omega_{LO})t}}{2} \qquad (2\text{-}5)$$

Equation (2-5) reveals how the quadrature mixer distinguishes between positive and negative frequencies. Frequencies above ω_{LO} (signal band) are down-converted to the right side of the spectrum, while frequencies below ω_{LO} (image band) are down-converted to the left side. Figure 2-21a depicts the ideal quadrature low-IF down-conversion. If the phase-shift between the local oscillator outputs is exactly 90° and their amplitudes are the same, no fraction of the image band is down-converted to the positive frequencies [1]. In the case of a phase-shift different of 90° and/or amplitude mismatch between LO_I and LO_Q, some power will leak from the image band into the signal band, and vice-versa (Figure 2-21b).

The image rejection ratio (*IRR*) quantifies the relation between the negative-half-plane image band and positive-half-plane leaked image power (Figure 2-21b) in quadrature receivers. The *IRR* can be expressed as a function of the finite mismatch between LO_I and LO_Q oscillator inputs, present in any analog implementation. This mismatch can be modelled as a gain error a_e and phase error ϕ_e. The effect of gain error in the complex mixing can then be calculated:

$$RF \times LO = IF = \cos(\omega_{RF}t) \cdot [(1+a_e) \cdot \cos(\omega_{LO}t) - j\sin(\omega_{LO}t)] \quad (2\text{-}6)$$

$$IF \approx \frac{1}{2}(e^{j(\omega_{RF}-\omega_{LO})t} + e^{j(-\omega_{RF}-\omega_{LO})t}) + \frac{a_e}{2}(e^{j(\omega_{RF}+\omega_{LO})t} + e^{j(\omega_{LO}-\omega_{RF})t}) \quad (2\text{-}7)$$

Equation (2-7) quantifies the fraction of the down-converted spectrum power that leaks from the positive-half-plane into the negative-half-plane, and vice-versa, due to the gain error a_e. The effect of the phase error ϕ_e in the complex mixing can be calculated as well:

$$RF \times LO = IF = \cos(\omega_{RF}t) \cdot [\cos(\omega_{LO}t) - j\sin(\omega_{LO}t + \phi_e)] \quad (2\text{-}8)$$

$$IF \approx \frac{1}{2}(e^{j(\omega_{RF}-\omega_{LO})t} + e^{j(-\omega_{RF}-\omega_{LO})t}) \frac{j\phi_e}{2}(e^{j(\omega_{RF}+\omega_{LO})t} + e^{j(\omega_{LO}-\omega_{RF})t}) \quad (2\text{-}9)$$

Equation (2-9) quantifies the fraction of the down-converted spectrum power that leaks from the positive-half-plane into the negative-half-plane, and vice-versa, due to the phase error ϕ_e. Equations (2-6) and (2-9) can be combined to give an approximate expression for the *IRR* [1]:

$$IRR_{dB} \approx 10 \cdot \log\left[\frac{4}{a_e^2 + \phi_e^2}\right] \tag{2-10}$$

Figure 2-22a depicts the zero-IF down-conversion in a quadrature receiver. The complex mixer output contains the wanted signal band shifted directly to dc, while the image band is up-converted to a higher negative band around $-2\omega_{LO}$. Because of their inherent simplicity, homodyne receivers are a very attractive solution for fully integrated receivers.

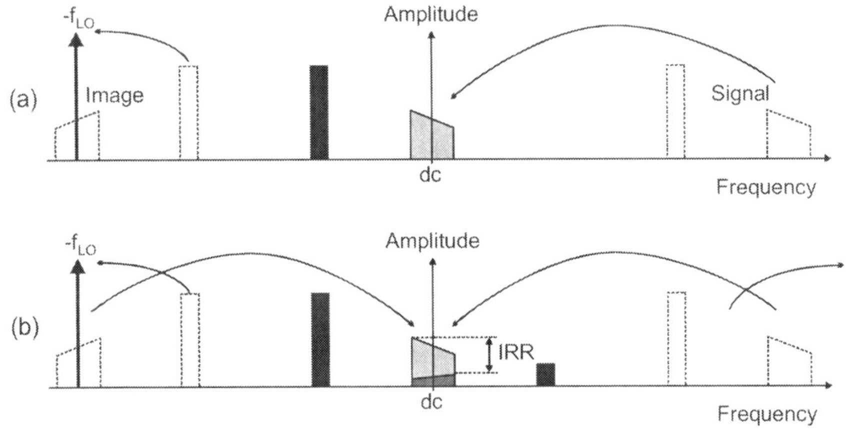

Figure 2-22: *Quadrature-paths zero-IF down-conversion with ideal matching (a), with I/Q mismatch (b).*

Figure 2-22b shows the effect of I/Q mismatch in the quadrature homodyne receiver. Because the image band is a mirrored version of the desired band for most modulation schemes, quadrature zero-IF receivers have much more relaxed image rejection requirements. The major drawback of the zero-IF architecture is that dc offset and $1/f$ noise can corrupt the baseband information. Furthermore, the fact that the mixer oscillator operates at same frequency as the IF may cause undesirable interferences [1].

2.5 Analog Radio Broadcasting

AM and FM are the most popular analog broadcasting radio services, being introduced, respectively, in 1920 and 1941 [25]. By far, the most widespread AM broadcasting band is the medium wave (MW), from 520 to 1,710kHz world wide. The channel width and the carrier spacing are 10kHz in the Americas, and 9kHz in the rest of the world. The long wave (LW) band, from 153 to 279kHz, has also been used in Europe, Africa and the Middle-East for AM broadcasting. The short wave (SW), from 2.3 to 26.1MHhz, has been used worldwide for very long distances broadcasting based on ionospheric reflection.

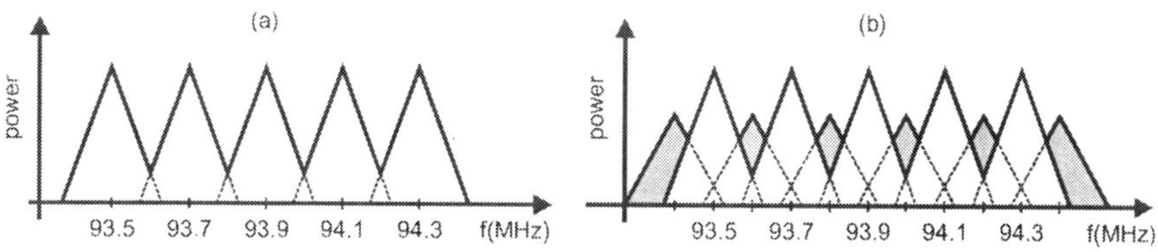

Figure 2-23: *FM channel distribution in the Americas (a) and in Europe (b).*

The FM radio broadcasting band ranges from 87.5 to 108.5MHz in most countries. An additional band, ranging from 65.9 to 74MHz, was also assigned in the former communist block, while Japan has its own FM radio allocation to the 76 to 90MHz band. In the Americas, the FM band carrier spacing and the modulated channel width are 200kHz (Figure 2-23a). In Europe, the carrier spacing is 100kHz and the channel width is 150kHz (Figure 2-23b). Because the FM band is overcrowded in Europe, adjacent FM channels are not used in the same region.

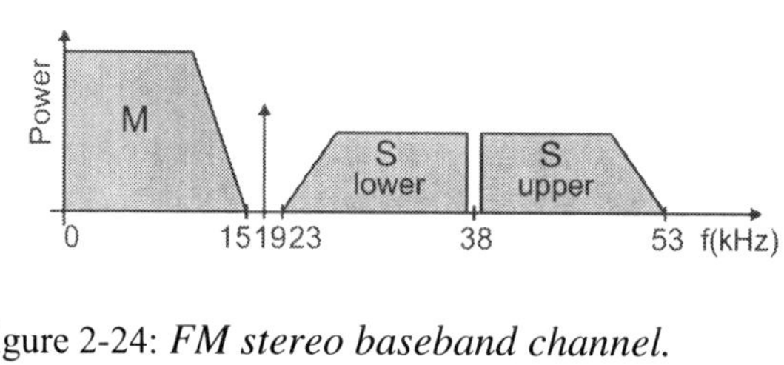

Figure 2-24: *FM stereo baseband channel.*

In order to accommodate stereo audio – left (L) and right (R) outputs – and maintain compatibility with mono receivers, a 55kHz FM baseband stereo channel consists of a mono compatible M=(L+R)/2 band and a double-sideband suppressed carrier S=(L-R)/2 band. The M band ranges from 30Hz to 15kHz, and a pilot tone is present at 19kHz. The S band, centred around 38kHz, ranges from 23 to 53kHz (Figure 2-24).

2.6 Digital Radio Broadcasting

In spite of the widespread use of AM/FM radios, the quality of the reception is limited by the characteristics of the old analog broadcasting standards. Several digital telecommunication techniques, developed since the 1980s, could have been employed for audio broadcasting. However, due to the lack of available spectrum, most of the proposals for terrestrial digital audio broadcasting require the shut down of the traditional analog transmission. As a result of the lack of backwards compatibility with AM/FM radio, in most of the world no full digital broadcasting solution has been adopted up to now for terrestrial broadcasting.

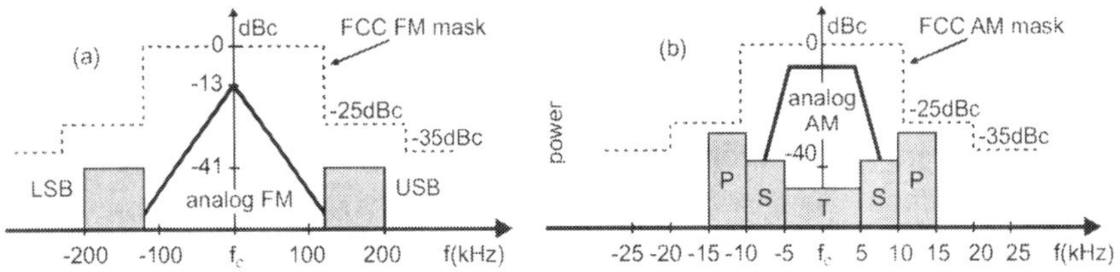

Figure 2-25: *Hybrid IBOC spectra. FM (a), AM (b).*

The *In-Band, On-Channel* (IBOC) standard, developed during the late 1990s, proposes a hybrid digital broadcasting solution within the traditional AM and FM bands [26]. The digital information is transmitted according to the FCC transmission masks for AM and FM broadcasting (Figure 2-25). Traditional receivers are able to receive the analog signal without noticeable quality loss on the reception because the digital broadcasting is perceived as an additional source of in-band noise. Due to the blend-to-analog feature, if the quality of the digital broadcasting is inferior to the analog broadcasting, an IBOC compatible receiver switches back to the analog reception mode.

A hybrid FM-IBOC channel is shown in Figure 2-25a. The upper and lower digital sidebands are present together with the analog information. The digital audio is transmitted using *Orthogonal Frequency Division Multiplexing* (OFDM) and *Forward Error Correction* (FEC) coding. Both digital sidebands contain the same baseband information to increase robustness against interferers from the 1st adjacent analog carrier, located 200kHz away from the tuned channel carrier. When both digital sidebands are properly received, because of the *Complementary Pair Convolution* (CPC) coding, additional audio quality improvement is achieved [27]. However, outside the Americas region, where the carrier distance is just 100kHz, FM-IBOC transmission is much less feasible.

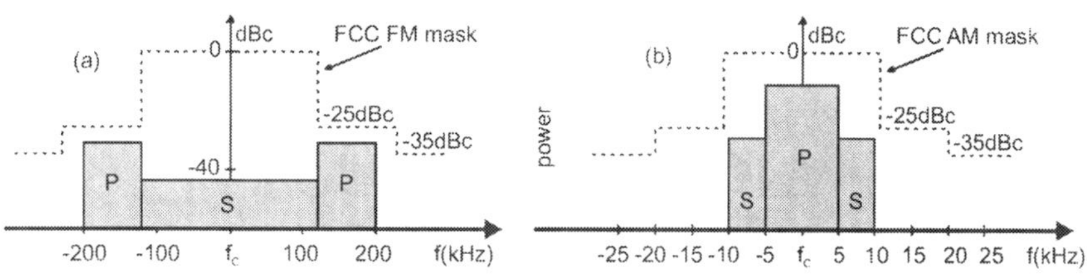

Figure 2-26: *All-digital FM (a) and AM (b) IBOC spectra.*

A hybrid AM-IBOC channel is shown in Figure 2-25b. Again OFDM and FEC coding are used for the primary (P) and secondary (S) digital audio sidebands. Because the analog information is amplitude modulated, some terciary (T) digital information is transmitted as a quadrature-phase component of the analog broadcasting. However, due to non-satisfactory performance caused by strong adjacent channel interferers, hybrid AM-IBOC is yet to be adopted [28].

Both FM and AM IBOC standards are prepared for transition to full digital broadcasting, if and when the analog broadcasting is shut-down. Figure 2-26 shows the all-digital FM and AM IBOC channel spectra. An alternative solution to increase the FM band digital broadcasting data rate, is to discontinue the stereo FM analog broadcasting. In this scenario, the stereo S channel bandwidth would be allocated to another pair of IBOC OFDM channels. The FM analog mono (M) channel service would continue available without interruption for backwards compatibility.

2.7 Integrated Solutions for AM/FM Receivers

Following the development of integrated circuit technology, several LSI (large scale of integration) based AM/FM receiver solutions became available by the mid-1970s [4]. This section presents an overview of commercial chip-sets available for car radios until the present time. Figure 2-27 shows a simplified diagram of a 4 ICs bipolar technology receiver chip-set. All active components required for AM demodulation were already integrated together. FM demodulation was partitioned between RF and IF ICs, while stereo decoding required its own dedicated analog processor. A 5th IC was employed in car radios to implement pulsed interference (e.g. from the ignition spark) cancellation. Several non-integrated components like filters, passives and crystal oscillators were also needed.

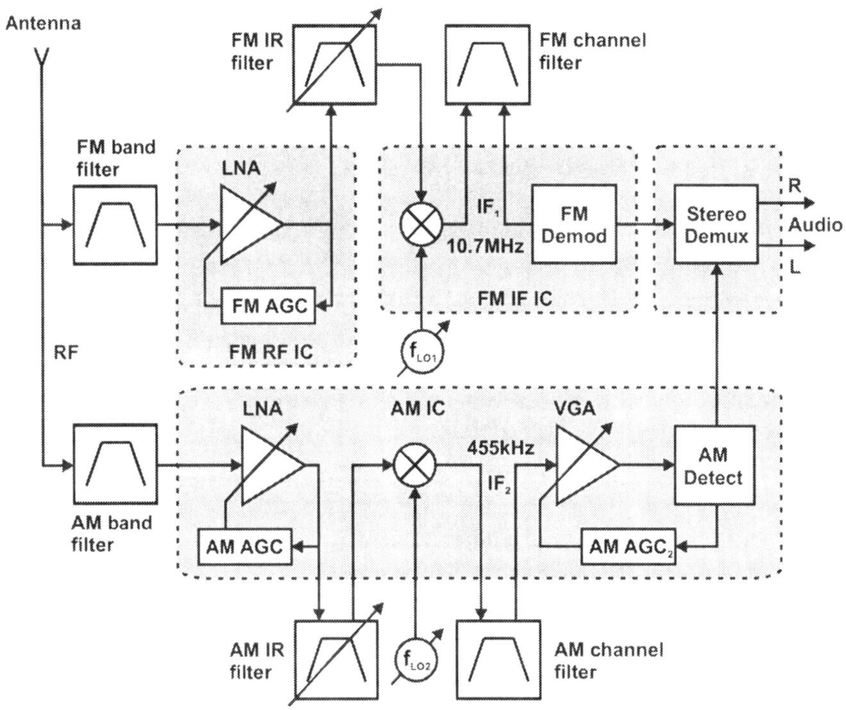

Figure 2-27: *Analog AM/FM receiver LSI chip-set [4].*

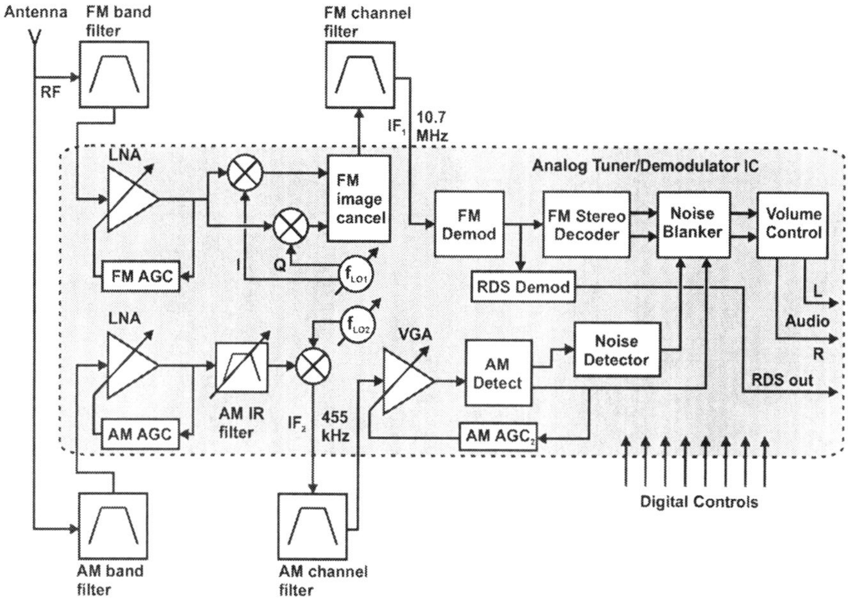

Figure 2-28: *Analog single-chip AM/FM receiver [7].*

In order to reduce manufacturing costs, the next generation of full analog radios was characterized by the integration of the receiver chip-set in a single bipolar IC for home radios [2], [3], car radios [4] and portable radios [5], [6]. Each specific application demanded a different emphasis on quality of the reception, immunity to interferers, lower power consumption and portability. A modern example of single-chip analog AM/FM receiver for car radios is [7]. The external IR FM filter is not needed anymore because of the quadrature FM image cancelling.

The advances on the digital CMOS technology during the 1980s allowed the implementation of several analog baseband functions, such as stereo deconding and audio control, in the digital domain [8]-[9]. A simplified diagram of a 2 ICs receiver is shown in Figure 2-29. The RF operations and demodulation are implemented in a single bipolar IC [11]. The audio analog outputs are digitised by a baseband ADC and in recent implementations ADC, DSP and DAC are integrated in single CMOS IC [10]. Due to the double-conversion architecture used for AM reception, an IR AM filter is not required.

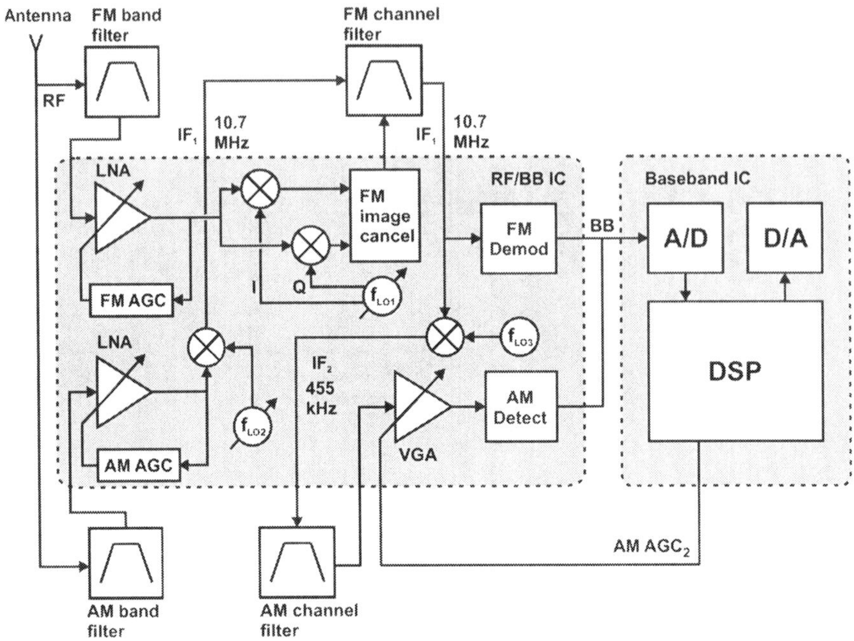

Figure 2-29: *AM/FM receiver chip-set with baseband ADC, DSP and analog demodulation [10]–[12].*

DSP based receivers have been a mainstream industry solution since the mid-1990s for AM/FM radios. An ADC operating at baseband has specifications relaxed by the several stages of mixing, filtering and amplification in the analog part of the receiver. If the analog-to-digital conversion takes place at the IF, the total amount of analog signal processing is reduced. Figure 2-30 shows an example of another 2 ICs radio receiver solution, with an ADC operating at the 10.7 MHz IF for both AM and FM signals. DSP based radios with IF ADCs also allow a substantial increase on the number of digital features available, as compared with previous radio generations [13], [14].

The first DSP radios with IF digitization were based on bandpass ΣΔ ADCs [13]–[18]. A more power efficient and linear solution is to use an IF-to-baseband ΣΔ ADC [20]–[21]. In order to achieve full channel selectivity, some channel filtering is implemented in the digital domain.

The radio receiver in Figure 2-30, built with a 82dB DR (200kHz BW) IF-to-baseband ADC [20], requires just one external 200kHz SAW filter with 40dB attenuation for FM. On the other hand, the full analog selectivity receiver in Figure 2-29 requires 3 similar filters in series to implement filtering with a higher quality-factor and more out-of-band attenuation [11]. Because of the high DR requirements of the AM reception (118dB in 3kHz BW), a programmable 0-18dB gain VGA has to be used in front of the 100dB DR (3kHz BW) IF-to-baseband ADC in the receiver shown in Figure 2-30.

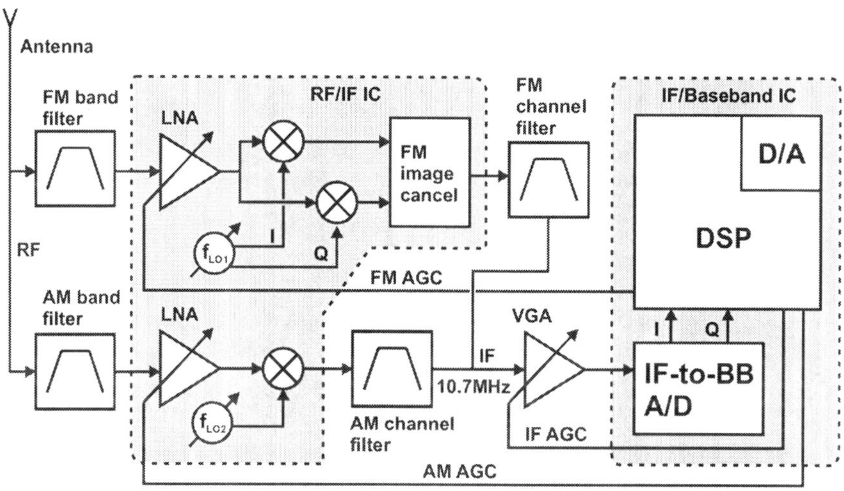

Figure 2-30: *AM/FM receiver chip-set with IF-to-baseband ADC and digital demodulation [20–21].*

In order to receive FM-IBOC channels, a DSP based radio requires two separate IF ADCs [21]–[22]. Figure 2-31 shows an FM-IBOC compliant commercial car radio receiver chip-set. An ADC and channel filters for AM and FM are required for the analog reception. Another ADC preceded by an FM-IBOC external channel filter (500kHz BW) is also connected to the FM front-end. Inside the DSP, the IBOC audio is demodulated and compared, in real-time, with the baseband audio received from the conventional FM channel. The IBOC algorithm then chooses the audio stream with the better quality. The reception of hybrid AM-IBOC channels does not require any change to the receiver hardware.

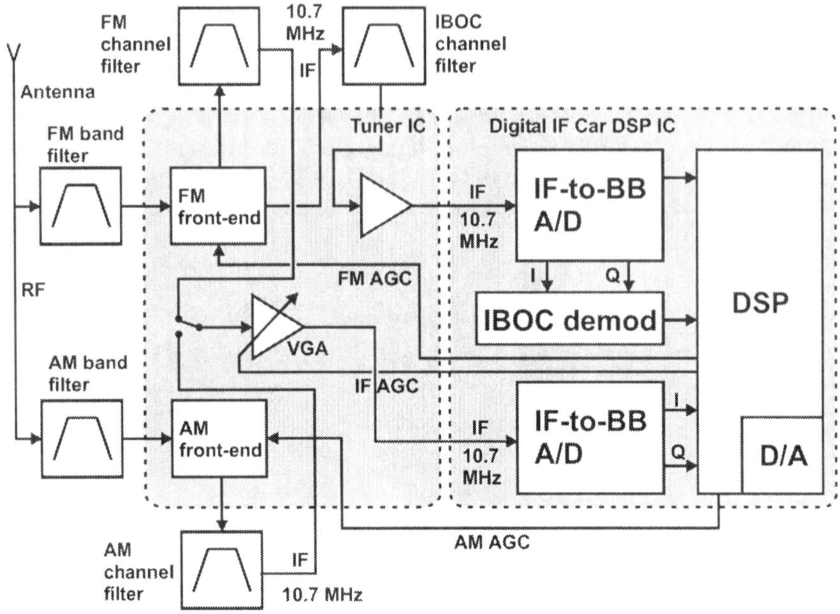

Figure 2-31: *AM/FM/IBOC receiver chip-set with IF-to-baseband ADCs and digital demodulation [21]–[22].*

To further reduce costs and improve performance, a 118dB DR ADC [23] can replace the combination of AM external channel filter, VGA and 100dB DR ADC. This receiver architecture (Figure 2-32) presents a better noise figure and reduced complexity. Since the AM filter presents a limited selectivity (30kHz BW centred at the 10.7MHz IF), the sensitivity against strong AM blocking channels is improved because all the required DR is provided by the IF ADC. The VGA, the AM filter and the IF AGC loop are eliminated from the receiver (section 2.3). In this case the ADC input contains 20 AM channels or a single FM channel, and most of the AM channel selection is performed in the digital domain. Due to the multi-channel AM digitization, a minimum 85dB IM_3 performance is also required from the high DR IF ADC. This ADC [23] consumes 210mW, about 8 times more than the current generation of IF-to-baseband ADCs [21]. However, the 230mW IF VGA is not part of the receiver anymore, and so the total power consumption of the proposed car radio chip-set in Figure 2-32 does not increase significantly.

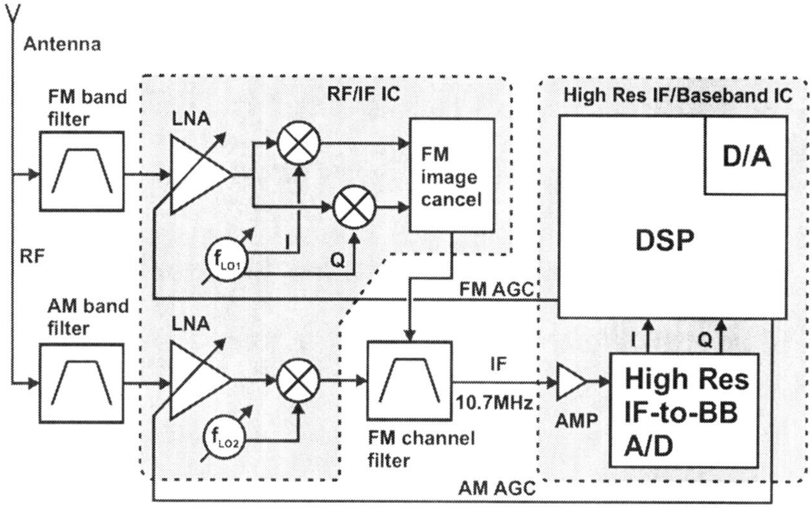

Figure 2-32: *Proposed AM/FM receiver chip-set with high-resolution IF-to-baseband ADC and digital demodulation [23].*

References

1. B. Razavi, "RF Microelectronics", *Prentice Hall, Upper Saddle River*, NJ, 1998.
2. W. Peil and R.J. McFadyen, "A Single Chip AM/FM Integrated Circuit Radio," *IEEE Trans. Consumer Electron.*, vol. 23, 424–429, Aug. 1977.
3. T. Okanobu, T. Tsuchiya, K. Abe and Y. Ueki, "A Complete Single Chip AM/FM Radio Integrated Circuit", *IEEE Trans. Consumer Electron.*, vol. 28, 393–407, Aug. 1982.
4. S. Sugayama, A. Kabashima, H. Suzuki, M. Yamagishi and T. Saeki, "Development of One-Chip IC for Car Tuners", *IEEE Trans. Consumer Electron.*, vol. 34, 642–648, Aug. 1988.
5. T. Okanobu, H. Tomiyama and H. Arimoto, "Advanced Low Voltage Single Chip Radio IC", *IEEE Trans. Consumer Electron.*, vol. 38, 465–475, Aug. 1992.
6. D. Yamazaki, C. Nishi and T. Okanobu, "A Complete Single Chip AM Stereo/FM Stereo Radio IC", *IEEE Trans. Consumer Electron.*, vol. 40, 563–569, Aug. 1994.

7. TEF6903 AM/FM single-chip analog tuner/processor, NXP's Car Entertainment Designer's Guide 2006/2007.
8. J.E. Haug, et al., "A DSP based stereo decoder for automotive radio", *SAE Technical Paper Series*, Feb.1990.
9. D.M. Funderburk and S. Park, "A Digital Receiver Design for AM Stereo Signals Using a General Purpose Digital Signal Processor", *IEEE Trans. Consumer Electron.*, vol. 40, 64–74, Feb. 1994.
10. H. Coumans, A. Turley, A. Koks and H. Schuurmans, "One-Chip Car-Radio DSP with High Analog Performance", *IEEE ICCE Digest Techinical Papers*, 1996.
11. K. Kianush and C.S. Vaucher, "A Global Car Radio IC with Inaudible Signal Quality Checks", *IEEE ISSCC Digest Techical Papers*, Feb. 1998.
12. TEF6862 AM/FM analog tuner and SAA7706H/SAA7709H baseband car radio DSP, NXP's Car Entertainment Designer's Guide 2006/2007.
13. L. Vogt, D. Brookshire, S. Lottholz and G. Zwiehoff, "A Two-Chip Digital Car Radio", *IEEE ISSCC Digest Techical Papers*, Feb. 1996.
14. J.W. Whikehart, "DSP-Based Radio with IF Processing", *SAE Technical Paper Series*, Jan. 2000.
15. F. Aducci et al., "A DSP-based digital IF AM/FM car-radio receiver", Proceedings of the 29th ESSCIRC Estoril, Portugal, Sept. 2003.
16. TDA7515 RF front-end for FM/AM car radios and TDA7580 FM/AM IF sampling processor, STMicroelectronics.
17. SDR510100 RF front-end IC and SDR530100 IF analog interface IC, Freescale *Symphony* digital radio chip-set.
18. S. Jantzi, R. Schreier and M. Snelgrove, "Bandpass Sigma-Delta Analog-to-Digital Conversion", *IEEE Trans. Circ. Syst.*, vol. 38, 1406–1409, Nov. 1991.
19. L.J. Breems, E.J. van der Zwan, E.C. Dijkmans and J.H. Huijsing, "A 1.8mW CMOS ΣΔ Modulator with Integrated Mixer for A/D Conversion of IF Signal", *IEEE J. Solid-State Circ.*, vol. 35, 468–475, Apr. 2000.
20. E.J. van der Zwan, K. Philips and C.A.A. Bastiaansen, "A 10.7-MHz IF-to-Baseband Sigma Delta A/D Conversion System for AM/FM Radio Receivers", *IEEE J. Solid-State Circ.*, vol. 35, 1810–1819, Dec. 2000.

21. Q. Sandifort, L.J. Breems, C. Dijkmans and H. Schuurmans, "IF-to-Digital Converter for FM/AM/IBOC Radio", Proceedings of the 29th European Solid-State Circuit Conference Estoril, Portugal, Sept. 2003, pp. 707–710.
22. TEF6730HW Digital IF car radio front-end and SAF7730HV dual IF car radio DSP, NXP's Car Entertainment Designer's Guide 2006/2007.
23. P. Silva, L. Breems, K. Makinwa, R. Roovers and J. Huijsing, "An IF-to-Baseband ΣΔ Modulator for AM/FM/IBOC Radio Receivers with a 118 dB Dynamic Range", *IEEE J. Solid-State Circ.*, vol. 42, 1076–1089, May 2007.
24. L. Breems and J.H. Huijsing, "Continuous-Time Sigma-Delta Modulation for A/D Conversion in Radio Receivers", *Kluwer Academic, Boston, MA,* 2001.
25. S. Haykin, "An Introduction to Analog and Digital Communications", *Wiley New York,* 1989.
26. B.W. Kroeger and P.J. Peyla, "Compatibility of FM Hybrid In-Band On-Channel (IBOC) System for Digital Audio Broadcasting", *IEEE Trans. Broadcast.*, vol. 43, 421–430, Dec. 1997.
27. B.W. Kroeger and D. Cammarata, "Robust Modem and Coding Techniques for FM Hybrid IBOC DAB", *IEEE Trans. Broadcast.*, vol. 43, 412–420, Dec. 1997.
28. J.R. Detweiler, "Conversion Requirements for AM & FM IBOC Transmission", *iBiquity Digital Corporation White Paper,* available on-line at www.ibiquity.com

Continuous-Time ΣΔ Modulation 3

ΣΔ modulators are non-linear feedback systems that enable the implementation of high-resolution analog-to-digital converters using a coarse quantizer. This is possible because the output of the modulator is characterized by low quantization noise power in a narrow bandwidth. A high-resolution Nyquist-rate digital output is obtained after digital filtering and decimation of the modulator output. The basic principles of operation of continuous-time ΣΔ modulators are described in section 3.1.

An efficient way to improve the resolution of a ΣΔ modulator is to increase the order of the modulator. Each increment of the modulator's order theoretically translates to an extra 20dB/dec attenuation of the quantization noise within the bandwidth of interest. However, high-order modulators can be unstable. The concept of stability for ΣΔ modulators is discussed in section 3.2

Single-bit ΣΔ modulators of any order are also prone to exhibit a tonal behaviour. In section 3.3, it is shown how circuit non-idealites, together with the presence of tones, may limit the actual performance of a ΣΔ ADC. Finally, section 3.4 reviews the effects of clock jitter on continuous-time ΣΔ modulators.

3.1 Basic Principles

The resolution of an analog-to-digital converter can be increased within a certain bandwidth of interest by means of oversampling, noise shaping and digital post-processing. These are the main operating principles used in ΣΔ analog-to-digital converters (ADCs). The ADC comprises a feedback loop called ΣΔ modulator and a digital filter/decimator. The ΣΔ modulator is based on a modification of the Δ modulator [1]. It was patented in 1960 by Cutler [2]. In 1962, Inose et al. [3] proposed the basic modulator architecture that is used today. At that time, CMOS technology was not mature enough to allow the implementation of an ADC based on the ΣΔ modulation principle. Most of the basic theoretical fundamentals were developed by J.C. Candy [4]–[6]. During the 1980s, several fully integrated implementations were published [7].

Figure 3-1 shows a generic block diagram of a single-loop continuous-time (CT) ΣΔ modulator [8]. It comprises a loop filter, an ADC in the feedforward path, and a DAC in the feedback path. The loop filter processes the difference between the analog input and the coded output. The output of the loop filter is sampled much faster than the required Nyquist-rate and quantized by the ADC. The DAC provides an analog equivalent of the output that can be subtracted from the input signal to generate the error signal. The simplest ΣΔ modulator uses a single integrator as a loop filter and a fast comparator as quantizer (single-bit ADC). The coded output is then called a bitstream.

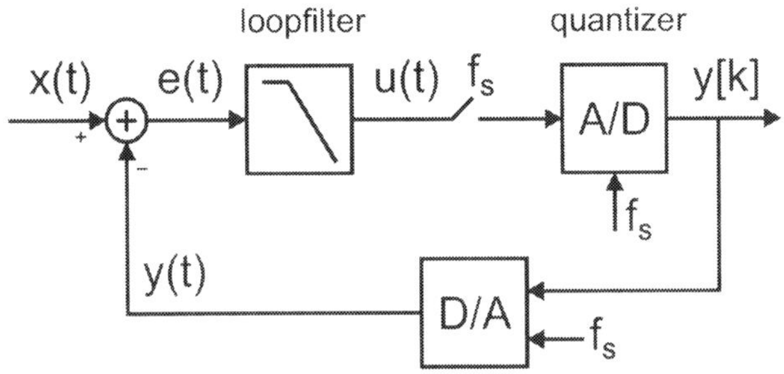

Figure 3-1: *Continuous-time ΣΔ modulator.*

Figure 3-2 shows the output bitstream and the sinusoidal input of a single-bit ΣΔ modulator. The effect of the quantizer on the system can be modelled as an addition of quantization error. In the case of a single-bit quantizer, the total power of the quantization error is very high inside the whole conversion bandwidth. However, the quantization error power at low frequencies is strongly attenuated. When combined with a digital decimator and lowpass filter, the ΣΔ modulator can provide very high-resolution analog-to-digital conversion.

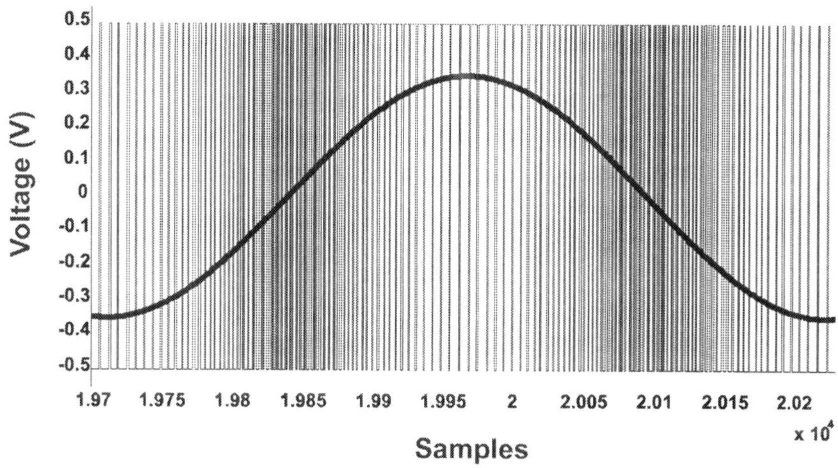

Figure 3-2: *Output bitstream and the sinusoidal input of a single-bit ΣΔ modulator.*

3.1.1 Oversampling

According to the *Nyquist theorem* [1], in order to sample an analog signal with limited bandwidth $[-f_B,+f_B]$ for error free reconstruction, the sampling frequency (f_S) should be at least two times larger than the single-sided bandwidth (f_B):

$$f_S \geq f_N = 2f_B \tag{3-1}$$

where f_N is defined as the Nyquist frequency. Figure 3-3 shows the generic output spectrum of a Nyquist-rate ADC.

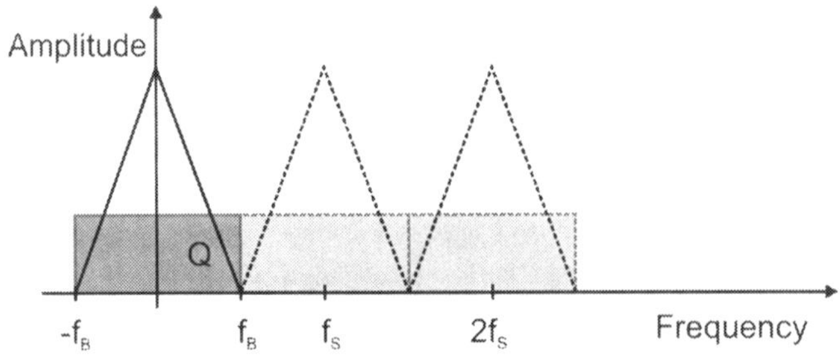

Figure 3-3: *Nyquist-rate ADC output spectrum (linear frequency scale).*

When the ADC operates at a rate greater than $2f_B$, the ADC is oversampled. In this case, the *oversampling ratio* (*OSR*) is the parameter that defines how many times f_S is larger than the minimum value required by the Nyquist theorem:

$$OSR = f_S / \ 2f_B \tag{3-2}$$

The output of an ADC contains information about the input signal and the quantization error. In case of a non-overloaded random input, the quantization error can be modelled as a white noise spectrum [1]. The power of the quantization noise is then a function of the quantization step size Δ:

$$q^2_{rms} = \frac{\Delta^2}{12} \tag{3-3}$$

Because of the sampling of the input signal, the quantization noise power is uniformly distributed inside the $[-f_S/2, f_S/2]$ bandwidth. The quantization noise PSD is then defined:

$$Q(f) = \frac{\Delta^2}{12f_S}, \left(\frac{-f_S}{f} < f < \frac{f_S}{2}\right) \tag{3-4}$$

The quantization noise power q^2_{rms} independs on the actual sampling frequency of the quantizer. This means that the higher the sampling frequency, the lower the quantization noise PSD. In this way the integral of the PSD $Q(f)$ remains constant from $-f_S/2$ to $f_S/2$. If the noise PSD is integrated in a fixed signal bandwidth $f_B < f_S/2$, the total noise power is smaller than q^2_{rms}. Therefore the integrated noise power within a fixed bandwidth decreases if the sampling rate increases. Mathematically, the in-band quantization noise power is given by:

$$n^2_{rms} = \int_{-f_B}^{f_B} Q(f)df = \frac{q^2_{rms}}{OSR} \tag{3-5}$$

Digital post-processing is required after the oversampled ADC in order to achieve a reduction of the quantization noise power. First, the coded output is filtered to eliminate the noise power outside the band of interest. Then the data is recoded and decimated to the Nyquist-rate. Figure 3-4 shows the generic output spectrum of an oversampled ADC.

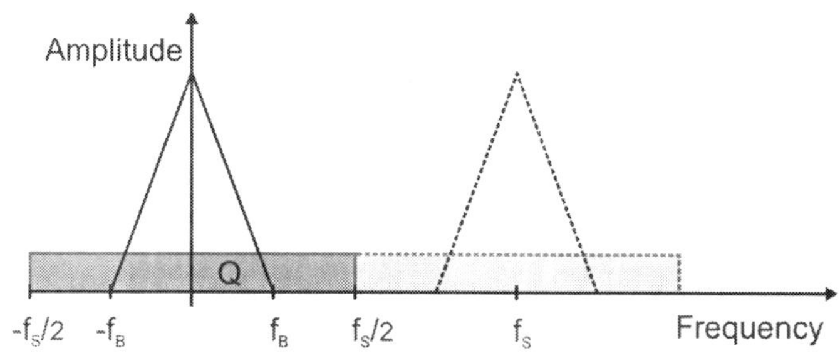

Figure 3-4: *Oversampled ADC output spectrum (linear frequency scale).*

The oversampling ratio can be rewritten as a power of 2: $OSR = 2^r$. The *SNR* of a Nyquist analog-to-digital converter with a maximum sinusoidal input, defined as been equal to the quantizer quantization step $A_{max}=\Delta$, can then be expressed as function of the number of bits (b) and the *OSR* [9.]:

$$SNR(dB) = 1.8 + 6b + 3r \tag{3-6}$$

Each time the *OSR* is doubled, the in-band *SNR* of a Nyquist ADC increases by 3 dB. Likewise, the effective resolution increases by half a bit after digital filtering of the out-of-band noise.

3.1.2 Noise Shaping

The concept of noise shaping is better understood after the analysis of the simplified linear model of a continuous-time ΣΔ modulator (Figure 3-5). The quantizer is modelled by an addition of quantization noise and the DAC is replaced by a linear unity gain feedback. The output *Y(s)* of the modulator can be calculated based on this model:

$$Y(s) = \frac{L(s)}{1 + L(s)} \cdot X(s) + \frac{1}{1 + L(s)} \cdot Q(s) \qquad (3\text{-}7)$$

where *X(s)* is the analog input, *L(s)* is the loop filter transfer function and *Q(s)* is the additive quantization noise PSD defined in Eq. (3-4). The term multiplying *X(s)* in Eq. (3-7) is defined as the *Signal Transfer Function* (*STF*). The *Noise Transfer Function* (*NTF*) is the term multiplying *Q(s)*. When *L(s)* is a lowpass filter with high dc gain, the *STF* has a value very close to 1 at low frequencies, while the *NTF* tends to zero. At low frequencies, the quantization error power is strongly attenuated. At high frequencies, the input signal is filtered and the quantization error is magnified. Figure 3-6 shows the generic output spectrum of a ΣΔ modulator, where oversampling and noise shaping are combined.

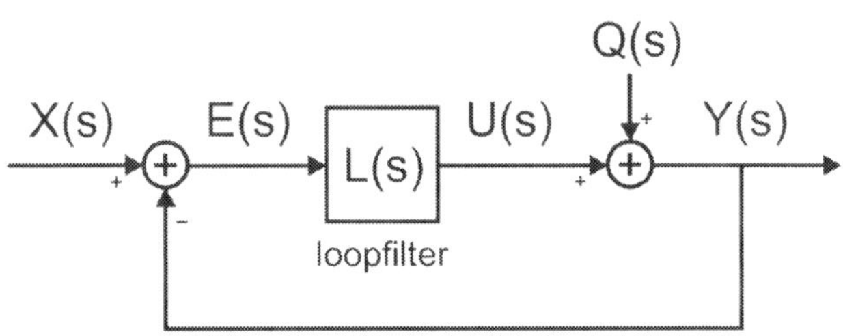

Figure 3-5: *Continuous-time ΣΔ modulator linear model.*

When the loop filter is implemented with one integration ($L(s)=\omega_S/s$), the ΣΔ modulator is of 1st-order. The theoretical peak *SNR* of any ΣΔ modulator can be calculated based on the linearised model of Eq. (3-7). For the 1st-order CT loop filter, the noise transfer function is:

$$NTF_1 = \frac{s}{s + \omega_S} \qquad (3\text{-}8)$$

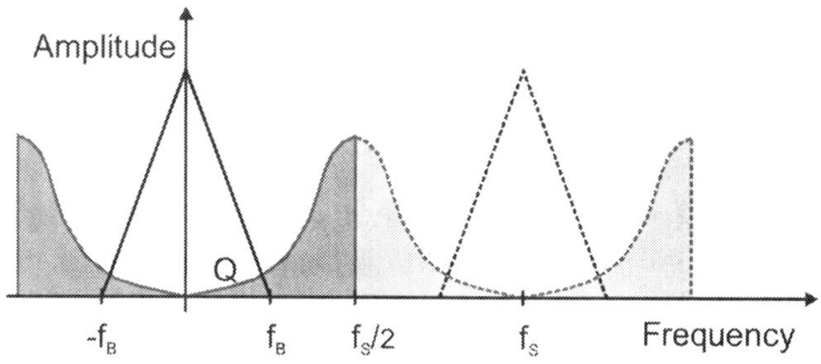

Figure 3-6: *Oversampled noise shaping ADC output spectrum (linear frequency scale).*

The noise power inside the band of interest is given by the integral of the shaped quantization noise PSD and it is obtained by combining Eqs. (3-4) and (3-8):

$$n^2_{rms} = \int_{-f_B}^{f_B} \left(\frac{\omega^2}{\omega^2 + \omega_S^2} \cdot \frac{q^2_{rms}}{f_S} \right) df \qquad (3\text{-}9)$$

The analytical solution of (3-9) is given by:

$$n^2_{rms} = 2q^2_{rms}\left(\frac{f_B}{f_S} - ArcTan\left(\frac{f_B}{f_S}\right)\right) \qquad (3\text{-}10)$$

The *ArcTan(x)* function can be approximated by its Taylor series expansion for $f_S >> f_B$:

$$ArcTan(x) = x - \frac{x^3}{3} + \frac{x^5}{5} - \ldots \qquad (3\text{-}11)$$

Replacing the expansion (3-11) in (3-10) and combining with (3-2) gives the final expression of the 1st-order shaped noise power:

$$n^2_{rms} = \frac{2}{3} \cdot \frac{q^2_{rms}}{OSR^3} \qquad (3\text{-}12)$$

If the modulator input signal is the maximum amplitude sinusoidal ($A_{max}=\Delta$), and the oversampling ratio is again expressed as $OSR = 2^r$, the peak SNR can be calculated:

$$SNR_{peak} = 9.03r \tag{3-13}$$

The effect of the noise shaping over the quantization noise is clear. In the case of pure oversampling, the in-band noise power reduction is linearly proportional to the OSR. When oversampling is combined with 1st-order noise shaping, the in-band noise reduction becomes proportional to OSR^3. Each time the OSR is doubled, the in-band SNR of the ΣΔ modulator increases by 9dB. Conversely, the effective resolution increases by 1.5 bits after digital filtering of the out-of-band noise.

3.1.3 Anti-alias Filtering

Generally, oversampled ADCs have much less stringent anti-aliasing requirements than Nyquist-rate ADCs [8]. However, switched-capacitor ΣΔ modulators implementations still require a low-order anti-aliasing filter. For this class of modulators, the sampling operation takes place before the input summing node. As shown in Figure 3-1, in continuous-time ΣΔ modulators, the sampling occurs after the loop filter. Therefore, the quantizer input signal is filtered and all sampling non-idealites inside the signal band are attenuated. This means that the continuous-time loop filter acts as an anti-alias filter [10]. The single-bit DAC in Figure 3-1 can be modelled as a sample-and-hold operator with the following transfer function:

$$D_H(f) = T \cdot Sinc(fT) \tag{3-14}$$

The substitution of (3-14) into (3-7) gives:

$$Y(f) = \frac{L(f)}{1 + L(f) \cdot D_H(f)} \cdot X(f) + \frac{1}{1 + L(f) \cdot D_H(f)} \cdot Q(f) \tag{3-15}$$

Because at low frequencies $L(f)$ has very high gain and the sample-and-hold function has roughly gain 1, inside the signal band Eq. (3-15) can be approximated by:

$$Y(f) \approx X(f) + Q(f) / L(f) \tag{3-16}$$

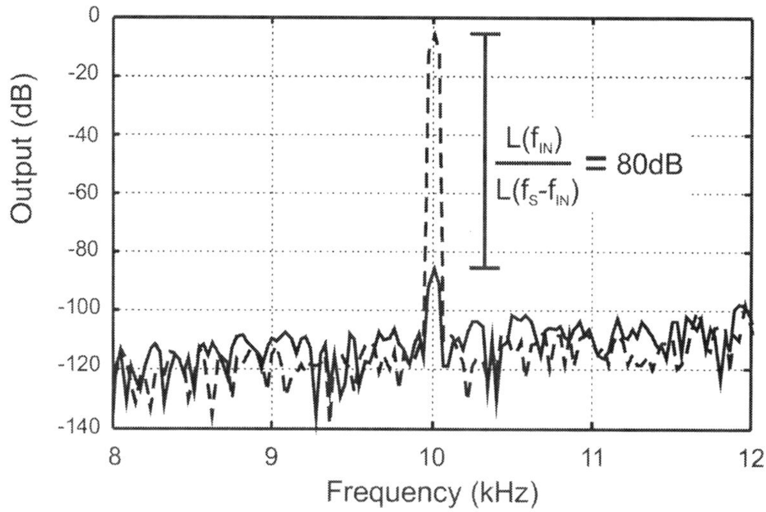

Figure 3-7: *Simulated low-frequency output spectrum of a 3rd-order CT ΣΔ modulator with -6 dB input at 10 kHz (dashed line) and at f_S-10kHz=1.99 MHz (solid line).*

According to the Nyquist theorem, the spectrum content of the loop filter output $U(f)$ around multiples of the sampling frequency f_S folds back to the low frequencies. The sampled version of $U(f)$ can then be expressed as [1]:

$$U_S(mf_S - f) = U(f) \tag{3-17}$$

Because the $sinc(fT)$ function is nulled at multiples of f_S, the spectral content of the feedback signal around multiples of f_S is strongly attenuated. Close to f_S, the loop filter output can be approximated as [10]:

$$U(f) \approx L(f) \cdot X(f) \tag{3-18}$$

Combining (3-16) and (3-18), the magnitude of the signal band aliased components in the modulator output spectrum can be estimated as:

$$Y(mf_S - f) \approx \frac{U(mf_S - f)}{L(mf_S - f)} \approx \frac{L(f)}{L(mf_S - f)} \cdot X(f) \tag{3-19}$$

Equation (3–19) explains how the aliased component of an input signal with spectral content above $f_S/2$ is suppressed by the continuous-time ΣΔ modulator. For instance, the aliased component in the frequency f_{IN}, caused by an input signal at f_S-f_{IN} is suppressed by the loop filter with a factor $L(f_{IN})/L(f_S\text{-}f_{IN})$. Figure 3-7 shows the simulated output spectrum of a continuous-time ΣΔ modulator. The input signal is a -6dB tone located at 10kHz (dashed line) or at 1.99MHz (f_S-10kHz). The tone in the solid line plot is the aliased component of the high-frequency input. It is clear that the aliasing attenuation factor is well predicted by (3-19):

$$L(10kHz)/\ L(1.99MHz)|_{dB} = 80dB \tag{3-20}$$

3.2 High-Order ΣΔ Modulators

In order to further increase the quantization noise attenuation in the band of interest, the order of the ΣΔ modulator loop filter can be increased. An intuitive way to implement a Nth-order modulator is by simply using a cascade of identical of integrators. The resulting loop filter can be generically expressed as:

$$L(s) = \left(\frac{\omega_I}{s}\right)^N \tag{3-21}$$

The equivalent *STF* and *NTF* can be calculated based on the linear model. At low frequencies, the *STF* is always an all-pass transfer function. The resulting Nth-order highpass *NTF* is given by:

$$NTF_N(s) = \frac{s^N}{s^N + \omega_I^N} \tag{3-22}$$

Figure 3-8 shows the magnitude Bode plot of several high-order *NTFs*. The linear model predicts an extra attenuation of 20dB/decade at low frequencies for every increment on the loop filter order (*N*). The theoretical in-band quantization noise reduction is then proportional to $OSR^{(2N+1)}$ [8]. Unfortunately, the extra noise shaping and resolution provided by high-order ΣΔ modulators can not be achieved by a simple cascade of integrators. The excessive loop filter phase delay (already 180° for a 2nd-order modulator) results in unstable modulators.

According to the linear systems theory, a dynamic system can be defined as stable, unstable or marginally stable. The states of a stable system are always bounded for a bounded input, and the states of an unstable system can be unbouded for a specific set of bounded inputs. A periodic solution of a dynamic system is named a limit cycle. Limit cycles can also be defined stable, semi-stable, or unstable [11]. In the case of a Nth-order ΣΔ modulator, the internal states are the output voltages of the *N* loop filter integrators. A ΣΔ modulator is then defined as stable when all its internal states are bounded for a bounded input.

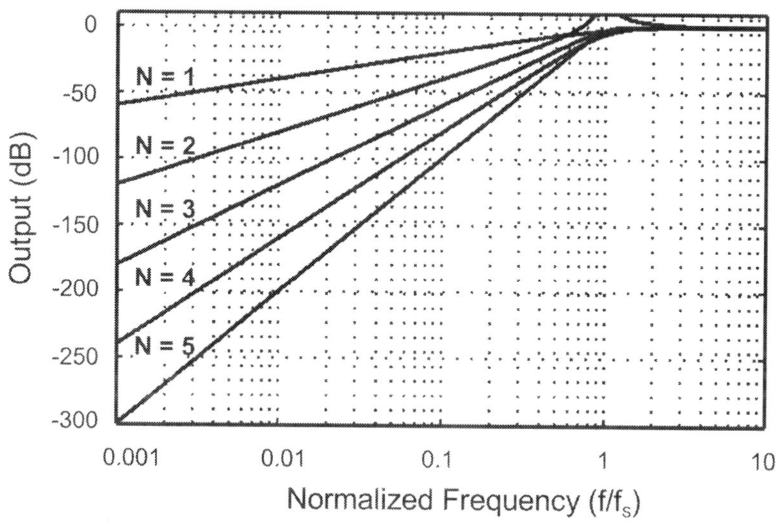

Figure 3-8: *Magnitude Bode plots for N^{th}-order CT NTFs (logaritmic frequency scale).*

The definition of instability of ΣΔ modulators based on lincar systems theory does not consider the fact that thc outputs of the loop filter integrators are always limited in a real implementation. Furthermore, the output bitstream and the internal states are stable limit cycles. A more suitable definition can be stated in terms of saturation of the internal states and frequency of the bitstream limit cycle. An internal state is said to be saturated when the output of the corresponding integrator is clipping to the positive and/or negative supply rail. A ΣΔ modulator is defined as being stable when its internal states are not saturated and the output bitstream presents a high-frequency limit cycle (nearby $f_S/2$). In this situation, the input signal information is preserved at low frequencies and high-resolution is achieved in the band of interest. A ΣΔ modulator is defined unstable when its internal states are saturated. The bitstream output then presents a lower frequency limit cycle ($f_S/8, f_S/12, f_S/16, f_S/24$ etc.) and the input signal information at very low frequencies is unrecoverable due to the high quantization noise floor.

Figure 3-9 shows the FFT of the output bitstream from a time-domain simulation of the 1b CT 2nd-order ΣΔ modulator with $L(s)=(\omega_S/s)^2$. The input signal is a -20dB sinusoidal at 10kHz. The reference level of the quantizer is 1 V and the clipping levels of the modulator are ±4 V. The *SNR* inside the 20kHz bandwidth (f_S=10MHz) is just 27dB, while the linear model (Eq. (3-22)) predicts a *SNR* of more than 90dB. The discrepancy between the linear model and the time-domain simulation is due to modulator unstable behaviour.

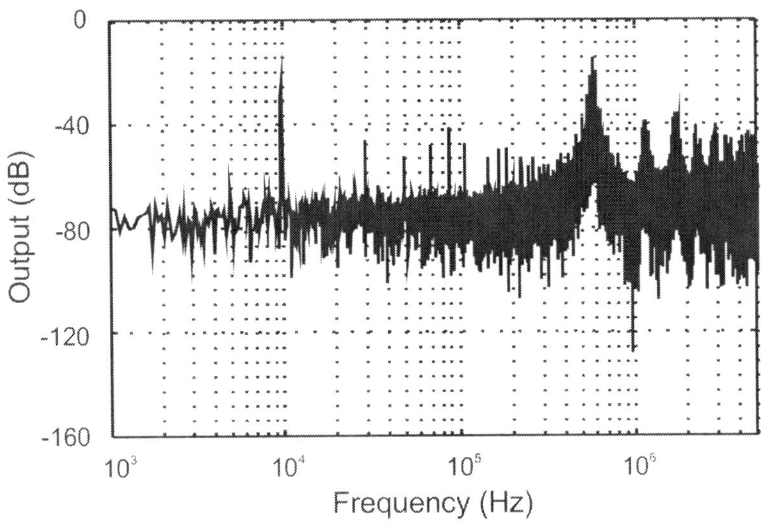

Figure 3-9: *Output spectrum of a 1b CT 2nd-order ΣΔ modulator with $L(s)=(\omega_S/s)^2$ (logarithmic frequency scale).*

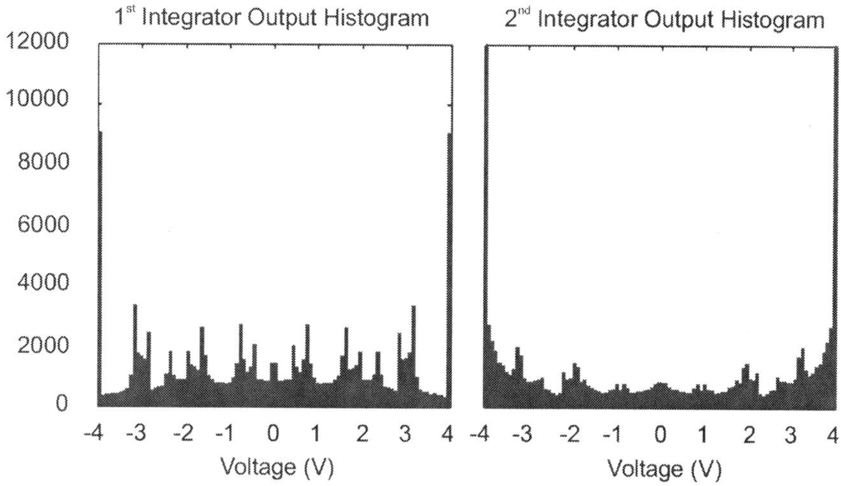

Figure 3-10: *Histogram of the internal states of a 1b CT 2nd-order ΣΔ modulator with $L(s)=(\omega_S/s)^2$.*

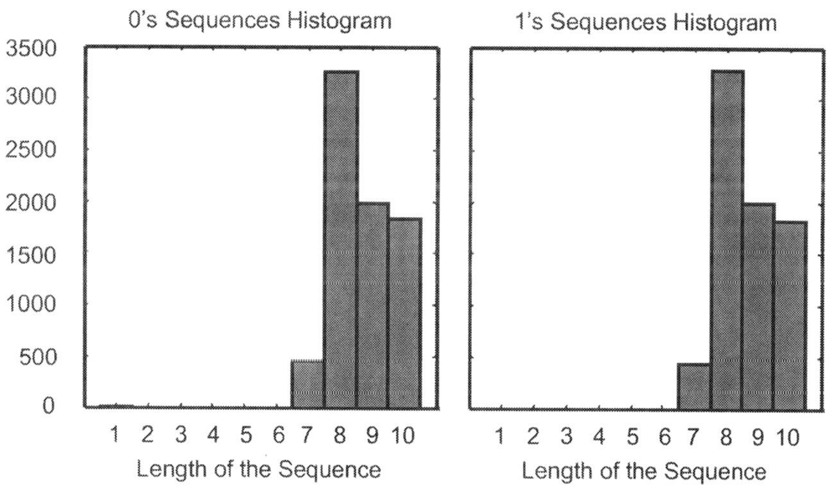

Figure 3-11: *Histogram of 0's and 1's sequences in the output bitstream of a 1b CT 2nd-order ΣΔ modulator with $L(s)=(\omega_S/s)^2$.*

The behaviour of the internal states of this unstable 2nd-order modulator is depicted in Figure 3-10. The histograms clearly show that the integrators' outputs are clipping. In this case the output bitstream is characterized by a low-frequency limit cycle. The histograms in Figure 3-11 show the lengths of the sequences of zeros and ones in the bitstream. The patterns with 8, 9 or 10 zeros followed by 8, 9 or 10 ones are the most common. Those patterns define a limit cycle oscillation between 500 and 625kHz that is clearly identified as the strongest tone in the output spectrum. In case of an idle input ideal simulation, the bitstream presents only sequences of 9 zeros and ones and the output FFT is exactly the spectrum of a square wave at $f_S/18$. The information about the ΣΔ modulator input appears in the output as a very low-frequency disturbance in the bitstream's "natural" limit cycle.

In order to stabilize the 2nd-order modulator, a zero can be introduced in the loop filter of Eq. (3-21). As a result the loop filter approaches a 1st-order system with -90° phase delay near $f_S/2$ while retaining the 2nd-order roll-off of -40dB/dec at low frequencies. High-order stabilization can be accomplished by means of feedforward, feedback, or any other architecture that allows the filter compensation with high-frequency zeros. Section 4.1 discuss in more detail several different architectures for stable high-order loop filters.

Figure 3-12 shows the FFT of the output bitstream from a time-domain simulation of a stabilized single-bit CT 2nd-order ΣΔ modulator with $L(s)=\omega_S(2s+\omega_S)/s^2$. The input signal is again a -20dB sinusoidal at 10kHz and the reference level of the quantizer is 1V. The SNR inside the 20kHz bandwidth (f_S=10MHz) is 92dB, in agreement with the linear model prediction shown in Figure 3-8, and the -40dB/dec noise shaping is clearly recognized.

The internal states of this single-bit CT 2nd-order modulator are shown in Figure 3-13. The integrators outputs are bounded to a differential voltage below the clipping level (±4V). This allows the modulator output to be composed of short sequences of zeros and ones (Figure 3-14). Most quantization noise power is concentrated in several tones between $f_S/2.5$ and $f_S/2$, and high-resolution analog-to-digital conversion is obtained at low frequencies.

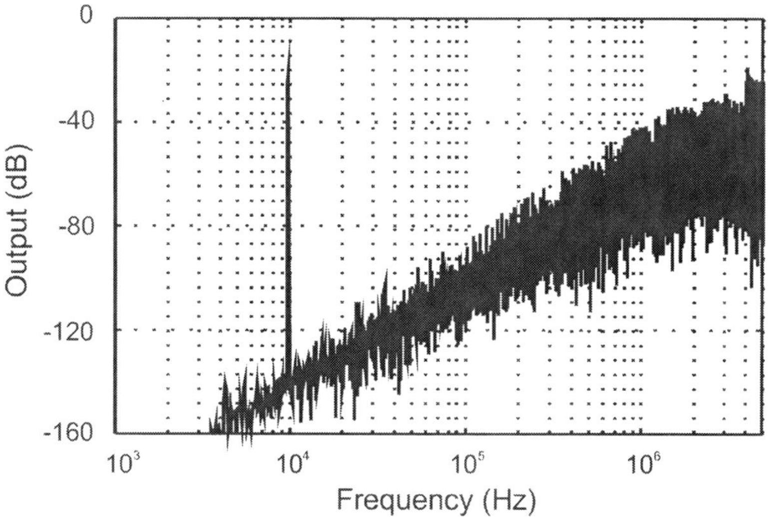

Figure 3-12: *Output spectrum of a 1b CT 2nd-order ΣΔ modulator with $L(s)=\omega_S(2s+\omega_S)/s^2$ (logarithmic frequency scale)*

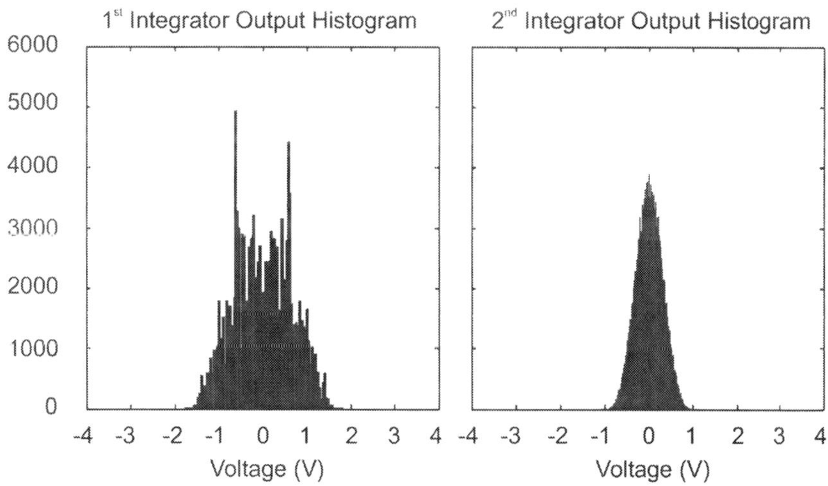

Figure 3-13: *Histogram of the internal states of a 1b CT 2nd-order ΣΔ modulator with $L(s)=\omega_S(2s+\omega_S)/s^2$.*

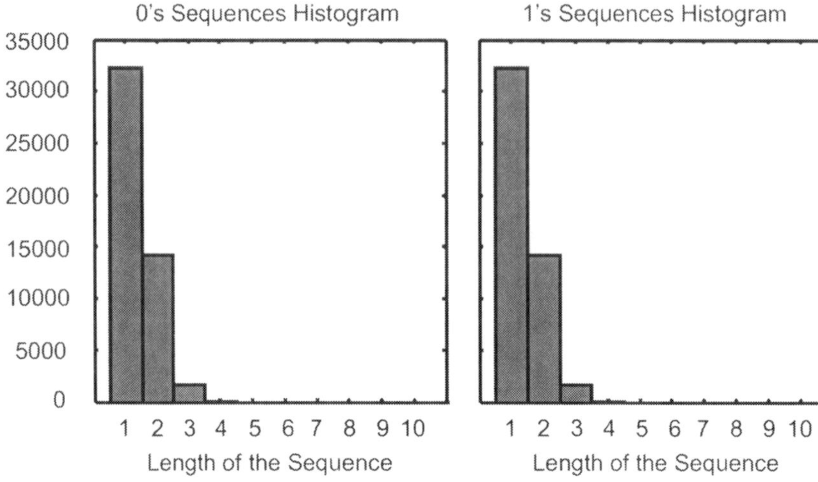

Figure 3-14: *Histogram of 0's and 1's sequences in the output bitstream of a 1b CT 2nd-order ΣΔ modulator with $L(s)=\omega_S(2s+\omega_S)/s^2$.*

The limit cycle analysis presented in this section gives some qualitative insight into the issue of stability of high-order ΣΔ modulators. A more quantitative relation between the loop filter coefficients and the modulator limit cycles can be derived based on the root locus technique. In [12] this analysis was developed for discrete-time loop filter transfer functions. The quantizer is modelled as a variable gain κ and a phase-shift θ, and the DAC is replaced by a zero-order holder. The modulator is stable if the poles of the non-linear characteristic equation of the closed-loop system are inside the unity circle. In [13] the method is developed for continuous-time transfer functions and successfully applied to the design of high-order low-pass ΣΔ modulators.

From a designer's point of view, the analysis presented in [12] is sufficient to guarantee the stability of high-order modulators with high-frequency zeros compensation. The design task then reduces to finding suitable continuous-time loop filters with 1st-order phase-shift near $f_S/2$ and enough quantization noise attenuation within the band of interest.

3.3 Tonal Behaviour

When the input of a multi-bit quantizer is a random signal, the quantization error can be safely approximated as a white noise process. In the case of a single-bit 1st-order ΣΔ modulator, however, the input and the output of the quantizer are rather correlated when the modulator input is a deterministic signal [8]. Figure 3-15 shows the output spectrum of a CT single-bit 1st-order ΣΔ modulator with a sinusoidal input. The quantization error is a periodic sequence whose power spectrum is concentrated at discrete tones, and is very far from being shaped white noise.

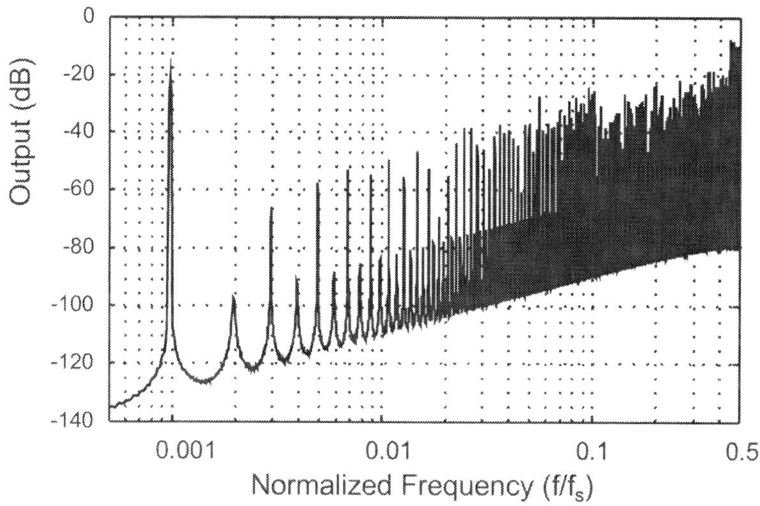

Figure 3-15: *Output spectrum of a 1b CT 1st-order ΣΔ modulator with sinusoidal input (logaritmic scale).*

In high-order ΣΔ modulators, the input of the quantizer is much less correlated to the modulator input. However, even high-order modulators are prone to tone generation for small dc inputs. In this case, the quantization error is again a periodic sequence. The frequencies of the tones are predicted by the empirical relation [8]:

$$f_n = nf_S \cdot \frac{|V_{dc}|}{2\Delta}, \quad n = \{1,\ 2,\ 3,\ ..\} \tag{3-23}$$

Figure 3-16 shows the detailed low-frequency output spectrum of a single-bit 3rd-order CT ΣΔ modulator. The dc input is 2.05 mV, f_S = 1 MHz and the quantization step (Δ) is 1 V. A -100 dB tone at 4.1 kHz (n = 4) and its harmonics (n = 8, 12, 16, 20) are clearly visible. Because the loop gain is smaller far from dc, the power of the tones increases with frequency. Likewise, tones cannot be seen at very low frequencies where the loop gain is very high. The power of baseband tones can be reduced if the loop gain is equally distributed inside the band of interest. This can be accomplished by spreading the *NTF* zeros in the signal band. High-order modulators with several *NTF* zeros in the band of interest are discussed in section 4.1.

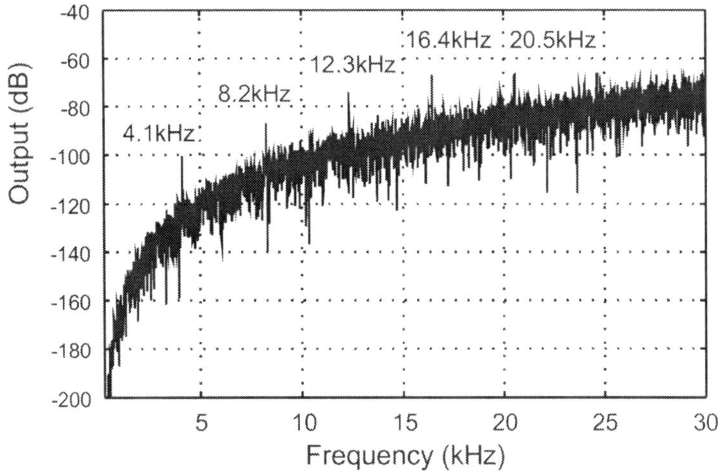

Figure 3-16: *Low-frequency output spectrum (linear frequency scale) of a 1b CT 3rd-order ΣΔ modulator for 2.05 mV dc input (Δ=1 V).*

Due to the natural limit cycle behaviour of stable ΣΔ modulators, very strong tones are present at about half the sampling frequency ($f_S/2$). Another empirical relation is used to calculate the location of these tones [8]:

$$f_n = \frac{f_S}{2}\left(1 - n\frac{|V_{dc}|}{\Delta}\right),\quad n = \{1,\ 2,\ 3,\ ..\} \qquad (3\text{-}24)$$

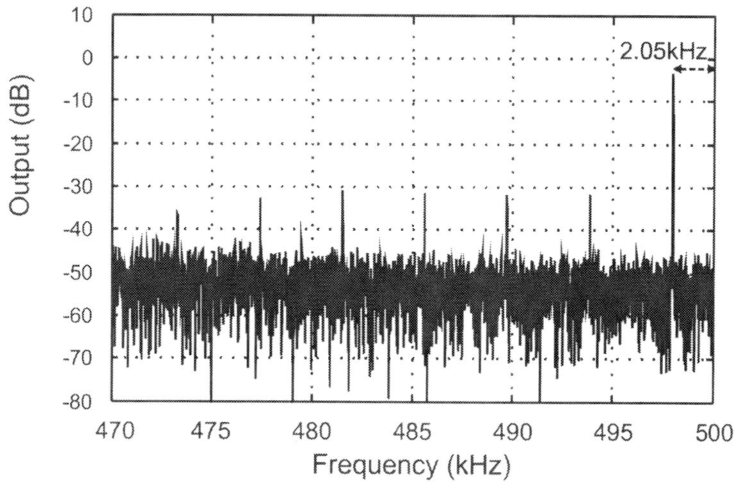

Figure 3-17: *High-frequency output spectrum (linear frequency scale) near $f_S/2$ of a 1b CT 3rd-order ΣΔ modulator for a 2.05mV dc input.*

Figure 3-17 shows the detailed output spectrum of the same single-bit 3rd-order CT ΣΔ modulator near $f_S/2$. A strong tone (-4dB) can be seen at 497.95kHz. This tone is located exactly 2.05kHz below $f_S/2$. It corresponds to $n = 2$ according to Eq. (3-24). In principle, the presence of strong tones near $f_S/2$ does not affect the performance of a ΣΔ ADC. The out-of-band noise power is filtered out by the digital post-processor. However, some non-idealites may cause the folding back of the strong tones (and the surrounding quantization noise floor) to the band of interest. Two mechanisms are potentially harmful: cross-talk with a signal at $f_S/2$ and 2nd-order harmonic distortion.

3.3.1 DAC Modulation at $f_S/2$

The parasitic coupling of a signal to the feedback DAC's reference voltage results in the mixing of the bitstream with the same signal. If the parasitic signal has a spectral component at $f_S/2$ (and/or its odd multiples), the down-conversion of the strong tones near $f_S/2$ to the signal band takes place in the modulator's feedback path. If C is a factor describing how much cross-talk couples to the DAC's reference voltage V_{ref}, and $Q[n]$ is the current output of the single-bit quantizer (-1 or +1), the DAC's output can be expressed as:

$$V_{dac} = V_{ref}\left(Q[n] + C \cdot \cos\left(\frac{f_S}{2}\right) \cdot Q[n]\right) \qquad (3\text{-}25)$$

Any bitstream component located at a frequency ($f_S/2$-f_n) with magnitude A is down-converted as a baseband tone:

$$BasebandTone = C \cdot A \cdot \cos(f_n) \qquad (3\text{-}26)$$

Figure 3-18 shows the low-frequency spectrum of the single-bit 3rd-order CT ΣΔ modulator with a 2.05mV dc input. The V_{ref} modulation coefficient is C=-80dB. As a result, the high-frequency tone located at 497.95kHz (Figure 3-17) is down-converted to 2.05kHz with -85dB power. Moreover, the -50dB quantization noise floor near $f_S/2$ is also down-converted, and the modulator *SNR* is reduced. The dc noise level is raised from less than -200dB to -130dB.

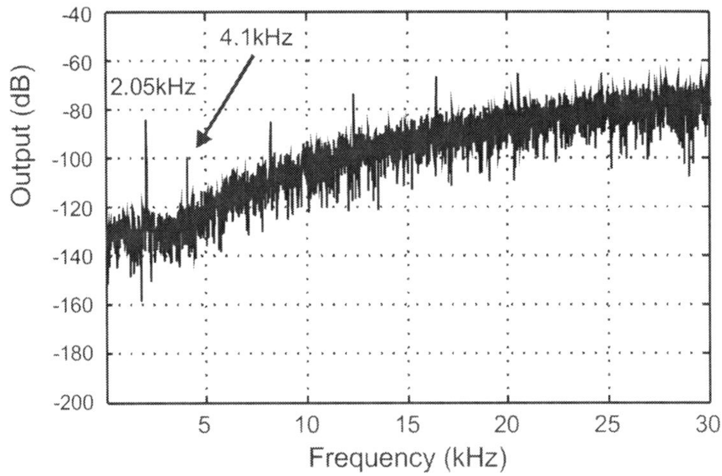

Figure 3-18: *Low-frequency output spectrum (linear frequency scale) of a 1b CT 3rd-order ΣΔ modulator for a 2.05mV dc input and cross-talk modulation at $f_S/2$ (C = -80dB).*

It is interesting to note that V_{ref} modulation at frequencies near f_S and its multiples are harmless to ΣΔ modulators. In this case, the power around f_S is also down-converted to baseband. However, the quantization noise in these frequencies is strongly attenuated by the modulator's noise shaping. No tones are mixed-down, neither can an increase of the in-band quantization noise be perceived.

3.3.2 2nd-Order Non-linearity

Assymetries in any fully-differential system can be perceived through the presence of even-order harmonic distortion, of which the 2nd-order distortion (HD_2) is the most harmful. Bitstream dependent loading of the DAC reference voltage also translates into 2nd-order distortion. The harmonics of an arbitrary bitstream spectral component ($f_S/2$-f_n) are located at:

$$f_{HD2} = f_S - 2f_n \tag{3-27}$$

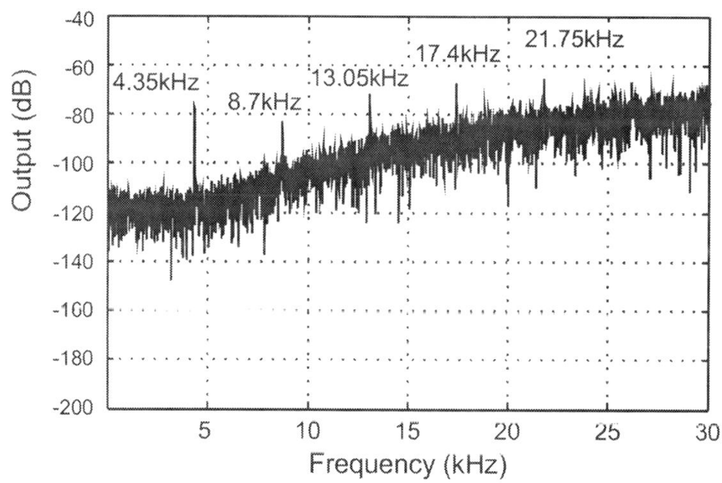

Figure 3-19: *Low-frequency output spectrum (linear frequency scale) of a 1b CT 3rd-order ΣΔ modulator with 2.17mV dc offset and -60dB memory effect on the feedback DAC reference.*

$$f_{HD3} = 3 \cdot \frac{f_S}{2} - 3f_n \qquad (3\text{-}28)$$

$$f_{HD4} = 2f_S - 4f_n \qquad (3\text{-}29)$$

In a CT ΣΔ modulator, sampling at a rate f_S takes place in the quantizer. According to the Nyquist theorem, any spectral component $f_X > f_S$ is aliased back to a frequency location $f_{AX} = | f_S - f_X |$. The harmonic components $f_{HD2}, f_{HD3}, f_{HD4}$ are then aliased to the following frequencies:

$$f_{A2} = 2f_n \qquad (3\text{-}30)$$

$$f_{A3} = \frac{f_S}{2} - 3f_n \qquad (3\text{-}31)$$

$$f_{A4} = f_S - 4f_n \qquad (3\text{-}32)$$

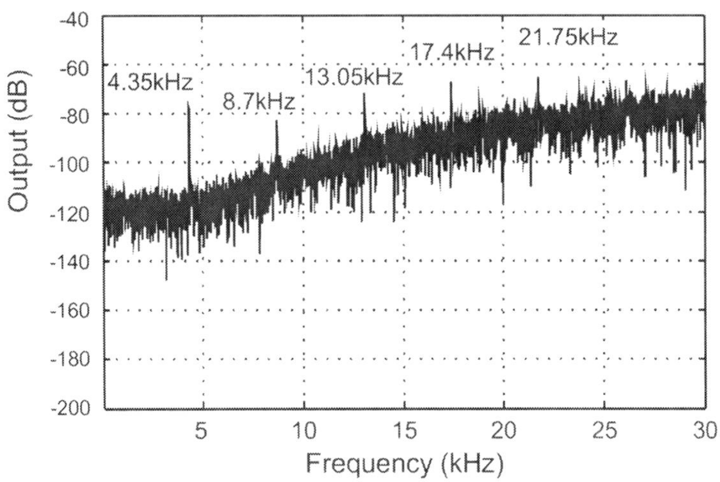

Figure 3-20: *Low-frequency output spectrum (linear frequency scale) of a 1b CT 3rd-order ΣΔ modulator with 2.17mV dc offset and -60dB memory effect on the feedback DAC reference.*

The aliased tones from the 3rd, 4th, etc. harmonics fall outside the frequencies of interest or have power levels that are too low to be perceived. However, the aliased tone f_{A2}, due to 2nd-order harmonic distortion, is located in the signal band if f_n is relatively small. Even-order harmonic distortion also causes a small change in the input dc level. As a consequence, the frequency location of the tones will move according to the change on the dc offset level.

Figure 3-20 shows the detailed spectrum at low frequencies of the single-bit 3rd-order CT ΣΔ modulator with a 2.05 mV dc input and -60 dB bitstream dependent loading of the feedback DAC reference voltage. The position of the baseband and high-frequency tones follow Eqs. (3-23) and (3-24). As a result of the increasing of the dc offset to 2.17 mV, the high-frequency tone located at 497.95 kHz in Figure 3-17 is shifted to 497.825 kHz. Due to 2nd-order harmonic distortion, this tone (and the nearby noise floor) is down-converted to 4.35 kHz. The frequency location of the original baseband tones is also slightly shifted (compare with Figure 3-16).

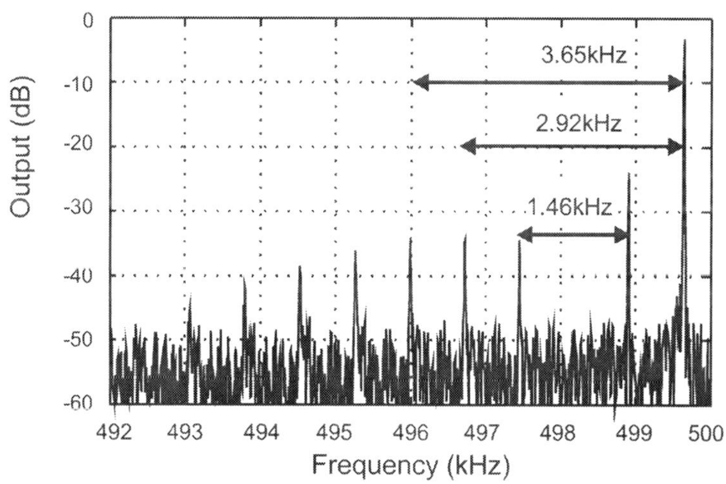

Figure 3-21: *High-frequency output spectrum (linear frequency scale) for a 0.73 mV dc offset and -60 dB DAC V_{ref} memory effect.*

The 2nd-order non-linearity also causes intermodulation distortion if two or more strong tones are present in the modulator. The 2nd-order intermodulation products (IM_2) of two tones (f_A, f_B) are located at f_A-f_B and f_A+f_B. If the tone frequencies f_A and f_B are close to each other, the *IM2* tone (f_A-f_B) falls inside the signal band [8]. The following example is very illustrative. The output of a single-bit CT 3rd-order ΣΔ modulator with 0.25 mV dc input presents no significant tones at baseband and several tones nearby $f_S/2$ (500kHz). The inclusion of -60dB bitstream dependent loading of the DAC reference voltage causes even-order non-linearity and changes the input offset to 0.73mV. The high-frequency tones are now located at f_1 = 499.64kHz, f_2 = 498.91kHz, f_3 = 497.45kHz, f_4 = 496.72kHz, f_5 = 495.99kHz and f_6 = 495.27kHz (Figure 3-21) with magnitudes of -3.3dB, -23.5dB, -35dB, -32dB, -35dB and -35dB. Figure 3-22 shows the detailed output spectrum at low frequencies for this modulator. The demodulated tones located at 1.46kHz (f_2-f_3), 2.92kHz (f_1-f_4), 3.65kHz (f_1-f_5) and 4.37kHz (f_1-f_6) are caused exclusively by intermodulation distortion. The tones located at 730Hz, 2.19kHz and 5.1kHz are caused by intermodulation distortion and aliasing of high-frequency HD_2 tones.

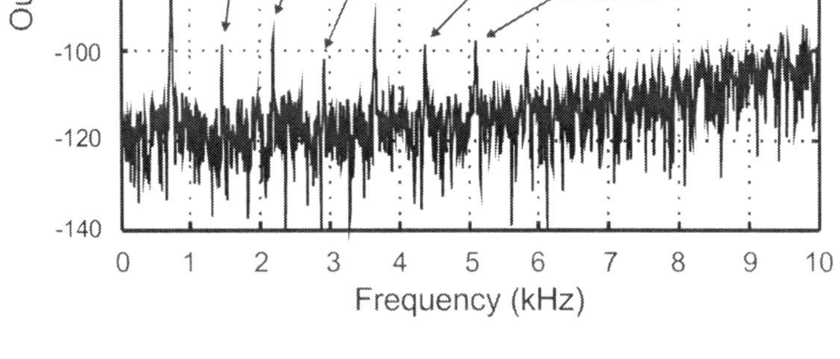

Figure 3-22: *Low-frequency output spectrum (linear frequency scale) for a 0.73mV dc offset and -60dB DAC V_{ref} memory effect.*

3.4 Clock Jitter

Continuous-time ΣΔ modulators (Figure 3-1) are affected by timing uncertainties in the quantizer during the sampling of the analog output of the loop filter, and in the DAC during the generation of the feedback pulses. The errors introduced during the sampling process are strongly attenuated by the modulator' s *NTF*. However, the errors introduced by the jitter of the DAC's feedback pulses are not attenuated, and may limit the performance of the entire modulator.

The magnitude of the jitter-induced errors depends on the pulse type that is fed back by the DAC. The most common DAC pulse shapes used in continuous-time ΣΔ modulators are: *Non-Return-to-Zero* (NRZ), *Return-to-Zero* (RTZ) and exponential. Figure 3-23 shows how these different DAC pulses are affected by timing uncertainties at the end of the pulse period. The grey areas portray the jitter-induced error charge. Because a feedback pulse is affected by jitter whenever a transition happens, NRZ DACs only introduce errors when the bitstream changes polarity. This is not the case for RTZ and SC DACs, which generate exponential DAC pulses. For these DACs, every feedback pulse contributes a rising and a falling transition, independent of the bitstream pattern. The jitter-induced errors in a modulator with NRZ DAC depends on the bitstream's statistical properties. However, because ΣΔ modulators with NRZ DAC also suffer from intersymbol interference [13], only modulators with RTZ and SC DACs are further analysed in this section.

The jitter-induced error in both RTZ and SC pulses can be estimated for two different jitter models: one in which the timing errors are assumed to be a white random process [10], [14], [15], and one in which the timing errors are assumed to have a deterministic variation [17]. For CT ΣΔ modulators with single-bit DACs, different results arise from these two assumptions.

3.4.1 Random Jitter

In this section, timing uncertainties are modelled as a random white noise process with variance σ_j^2. In Figure 3-23, timing errors Δt cause a random modulation of the pulse width. The effect of pulse-width jitter on the *SNR* of a CT ΣΔ modulator can be calculated for RTZ and SC pulses [14].

RTZ feedback

The variance of the integrated error charge σ_q^2 per clock cycle T_s, for a RTZ pulse, can be expressed by [14]:

$$\sigma_q^2 = \frac{\sigma_j^2 \cdot I_{DAC}^2}{\delta} \tag{3-33}$$

Equation (3–33) is derived taking into account the fact that clock jitter at the output of an oscillator depends on its output frequency [14]. The *SNR* related to pulse-width jitter errors can be calculated for a sinusoidal input with a maximum amplitude A_{max}. Assuming that the DAC current is 3 dB above the maximum input amplitude ($I_{DAC} = A_{max} \cdot \sqrt{2}$), the peak *SNR* is [14]:

$$SNR_{RTZ} = 10\log\left(\frac{\delta \cdot OSR}{4 \cdot f_S^2 \cdot \sigma_j^2}\right) \tag{3-34}$$

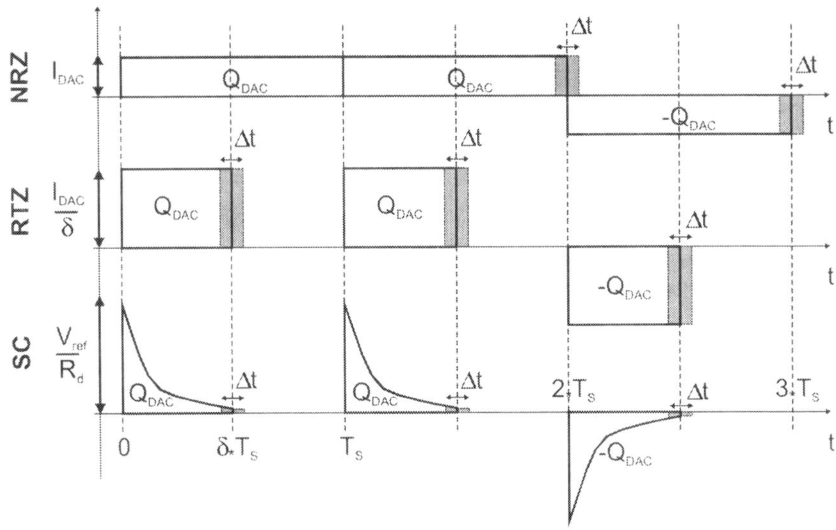

Figure 3-23: *Jittered feedback waveforms from NRZ, RTZ and SC single-bit DACs for the bitstream pattern 110.*

SC feedback

In the case of SC feedback, the error charge also depends on the DAC current I_{DAC} settling behaviour. The variance of the error charge σ_q^2 integrated per clock cycle T_S, for the exponential pulse in Figure 3-23, can be approximated by ($\Delta t << T_S$) [14]:

$$\sigma_q^2 = \sigma_j^2 \cdot \delta \cdot \left(\frac{V_{ref}}{R_d} \cdot e^{\frac{-T_S \cdot \delta}{\tau}}\right)^2 \tag{3-35}$$

where $\tau = R_d \cdot C_d$ is the time-constant of the DAC exponential current, V_{ref} is the voltage stored in the SC DAC capacitor C_d, and R_d is the equivalent series resistance in the discharging path. Assuming again a sinusoidal input with maximum amplitude A_{max}, the *SNR* limitation due to pulse-width jitter can be calculated [14]:

$$SNR_{SC} = 10\log\left[\frac{\delta \cdot OSR}{4 \cdot f_S^2 \cdot \sigma_j^2}\left(\frac{1-e^{\delta \cdot \alpha}}{\delta \cdot \alpha}\right)^2\right] \tag{3-36}$$

where $\alpha = T_S / \tau$ is defined as the relative time-constant inside the period T_S. The $(1-e^{\delta\alpha})/\delta\alpha$ term in (3-36) is the jitter sensitivity improvement factor from a RTZ to a SC DAC implementation.

Pulse-positioning jitter

Figure 3-23 describes how timing errors affect the width of different DAC pulses. The starting instant of the feedback pulses is also shifted in time due to clock jitter. The effect of pulse-positioning jitter on the ΣΔ modulator *SNR* is the same for any kind of pulse shape. The *SNR* degradation due to this effect is expressed by [10]:

$$SNR_{PulsePosition} = 10\log\left(\frac{1}{4 \cdot \pi^2 \cdot f_b^2 \cdot \sigma_j^2}\right) \tag{3-37}$$

where f_b is the edge of the signal band. According to the random jitter model, the *SNR* degradation due to pulse-positioning is the ultimate limiting factor due to timing uncertainties.

3.4.2 Deterministic Jitter

The random jitter model discussed in the previous section is not able to predict spurious tones in the ΣΔ modulator output spectrum due to a jittered DAC clock. This happens because the random jitter model does not account for the presence of discrete tones in the spectrum of real clock signals (Figure 3-24). These tones originate from unwanted coupling of spurious signals to the output node of a PLL, or from the modulation of the local oscillator by deterministic low-frequency signals [18].

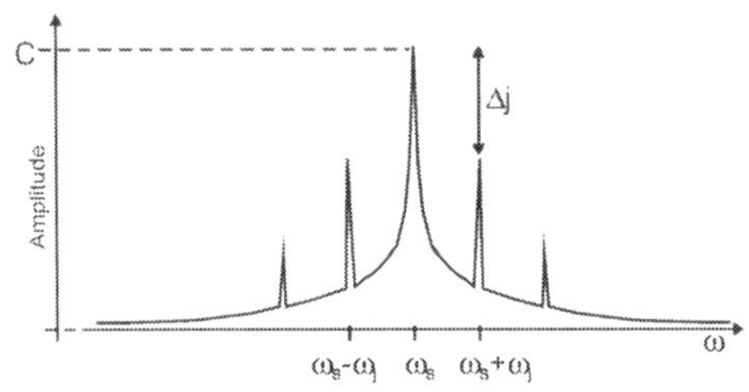

Figure 3-24: *Typical clock spectrum with spurious tones around the fundamental oscillation frequency.*

The spurious tones in Figure 3-24 can be modelled as the modulation of the clock fundamental component by a low-frequency sinusoide. Therefore, the clock signal is frequency modulated (FM) by a sine wave with time error amplitude Δt and frequency ω_j. The fundamental component of the jittered clock can be expressed by [16], [17]:

$$C_F(t) = C \cdot \cos(\omega_S \cdot t + \omega_S \cdot \Delta t \cdot \sin(\omega_j \cdot t)) \qquad (3\text{-}38)$$

where C is the amplitude of the clock fundamental component.

SC feedback

For low modulation frequencies ($\omega_j \ll \omega_S$), Figure 3-25 shows a spectral component in the ΣΔ modulator output and the jitter-induced sidebands. According to [17], the attenuation factor $\Delta s_{+/-}$ of each sideband is:

$$\Delta s_{+/-} = (\omega_S \cdot \Delta j) / (\omega_y +/- \omega_j) \qquad (3\text{-}39)$$

where $\Delta j = 2/(\omega_s . \Delta t)$ is the distance between the clock fundamental frequency and the jitter-induced adjacent tones.

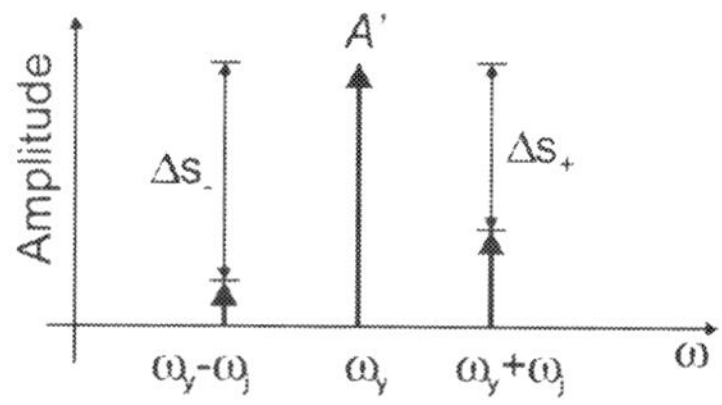

Figure 3-25: *Effect of the sinusoidal jitter on each spectral component in the output of a CT ΣΔ modulator with single-bit SC DAC.*

RTZ feedback

In the case of RTZ DAC, both jitter-induced sidebands have the same attenuation factor $\Delta s_{+/-}$ [17]:

$$\Delta s_{+/-} = \frac{\omega_s \cdot \Delta j}{\omega_y} \tag{3-40}$$

References

1. S. Haykin, "An Introduction to Analog and Digital Communications", *Wiley, New York*, 1989.
2. C. Cutler, "Transmission systems employing quantization", *U.S Patent 2.927.962*, Mar. 1960.
3. H. Inose, Y. Yasuda and J. Murakami, "A Telemetering System by Code ΔΣ Modulation", *IRE Trans. Space Electron. Telemetry*, vol. SET-8, 204–209, Sept. 1960.
4. J.C. Candy, "A Use of Limit Cycle Oscillations To Obtain Robust Analog-to-Digital Converters", *IEEE Trans. Commun.*, vol. 22, 298–305, Mar. 1974.
5. J.C. Candy, "A Use of Double Integration in Sigma Delta Modulation", *IEEE Trans. Commun.*, vol. 33, 249–258, Mar. 1985.

6. J.C. Candy, "Decimation for Sigma Delta modulation", *IEEE Trans. Commun.*, vol. 34, 77–81, Jan. 1986.
7. J.C. Candy and G.C. Temes, "Oversampling Delta-Sigma Data Converters", *IEEE Press*, *New York*, 1992.
8. S.R. Norsworthy, R. Schreier and G.C. Temes, "Delta-Sigma Data Converters: Theory, Design and Simulation", *IEEE Press*, *Piscataway*, *NJ*, 1997.
9. P.M. Aziz, H.V. Sorensen and J. van der Spiegel, "An Overview of Sigma-Delta Converters", *IEEE Signal Proc. Mag.*, vol. 13, issue 1, Jan. 1996.
10. E.J. van der Zwan and E.C. Dijkmans, "A 0.2-mW CMOS ΣΔ Modulator for Speech Coding with 80dB Dynamic Range", *IEEE J. Solid-State Circ.*, vol. 31, 1873–1880, Dec. 1996.
11. W.E. Boyce and R.C. DiPrima, "Elementary Differential Equations and Boundary Value Problems", *Wiley, New York*, 1992.
12. J. van Engelen and R. van de Plassche, "Bandpass Sigma Delta Modulators", *Kluwer Academic*, *Boston, MA,* 1999.
13. L. Breems and J.H. Huising, "Continuous-Time Sigma-Delta Modulation for A/D Conversion in Radio Receivers", *Kluwer Academic*, *Boston, MA*, 2002.
14. R.H.M. van Veldhoven, "A Triple-Mode Continuous-Time ΣΔ Modulator with Switched-Capacitor Feedback DAC for a GSM-EDGE/CDMA2000/UMTS Receiver", *IEEE J. Solid-State Circ.*, vol. 38, 2069–2076, Dec. 2003.
15. M. Ortmanns, F. Gerfers and Y. Manoli, "A Continuous-Time ΣΔ Modulator with Reduced Sensitivity to Clock Jitter Through SCR Feedback", *IEEE Trans. CAS-I.*, vol. 52, 875–884, May 2005.
16. S. Harris, "The Effects of Sampling Clock Jitter on Nyquist Sampling Analog-to-Digital Converters and on Oversampling Delta-Sigma ADCs", *J. Audio Eng. Soc.*, vol. 38, July/Aug. 1990.
17. R.H.M. van Veldhoven, Peter Nuijten and Paul van Zeijl, "The Effect of Clock Jitter on the DR of ΣΔ Modulators, Proceedings of the IEEE ISCAS, Kos, Greece, May 2006.
18. B. Razavi, "A Study of Phase Noise in CMOS Oscillators", *IEEE J. Solid-State Circ.*, vol. 31, 331–343, Mar. 1996.

ΣΔ *ADC Topologies for Radio Receivers* 4

Several DSP based radio receiver architectures were described in Chapter 2 of this book. In this type of receiver, the ADC is one of the most important analog blocks. Specific characteristics of each combination of telecommunication standard and receiver architecture (e.g. single-path or quadrature-paths, baseband or IF digitization, etc.) dictate the choice of the most suitable ADC topology. Due to the high DR required by AM/FM standards, ΣΔ modulators have been the preferable solution for IF digitization in radio receivers.

This chapter discusses in detail five continuous-time (CT) ΣΔ ADC topologies. Section 4.1 deals with lowpass ΣΔ ADCs for baseband digitization, with emphasis on several compensation techniques for stable high-order loop filter design. Section 4.2 introduces IF-to-baseband ΣΔ ADCs, where an analog mixer is integrated with a lowpass ΣΔ modulator for IF digitization. Quadrature IF-to-baseband ΣΔ ADCs are the subject of section 4.3. Attention is given to the design of complex loop filters and the issue of image rejection in non-ideal complex systems. Bandpass and quadrature bandpass ΣΔ ADCs for IF digitization are analysed in sections 4.4 and 4.5. Section 4.6 finally presents a discussion about the most suitable IF ADC topology for the car radio application.

4.1 Lowpass ΣΔ ADCs

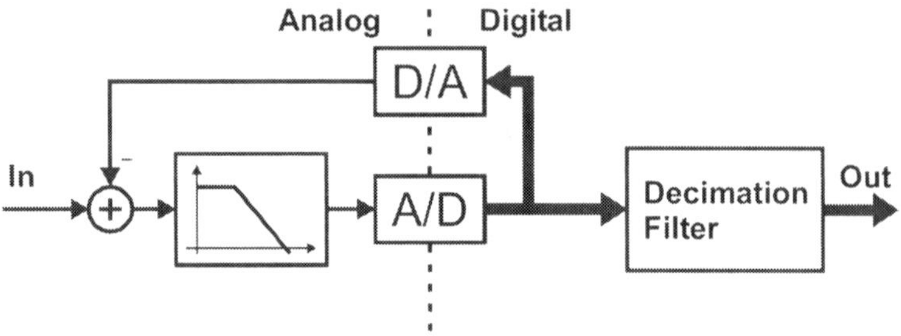

Figure 4-1: *Lowpass ΣΔ ADC for baseband digitization.*

Figure 4-1 shows the block diagram of a lowpass ΣΔ ADC. It consists of a ΣΔ modulator with a lowpass loop filter and a decimation filter. Both the input of the analog modulator and the digital output of the decimation filter are baseband signals. Figure 2-4 shows an example of a DSP based receiver with baseband digitization. The preceding stages of analog signal processing relax the ADC's resolution and speed requirements (section 2.1.2). This translates into relatively low power consumption when compared to other solutions [1].

Stable high-order lowpass ΣΔ ADCs can be designed based on several different loop filter topologies. Figure 4-2 shows a linear model of ΣΔ modulator with a generic loop filter [2]. The *STF(s)* and the *NTF(s)* can be expressed as a function of the loop filter transfer functions $L_0(s)$ and $L_1(s)$.

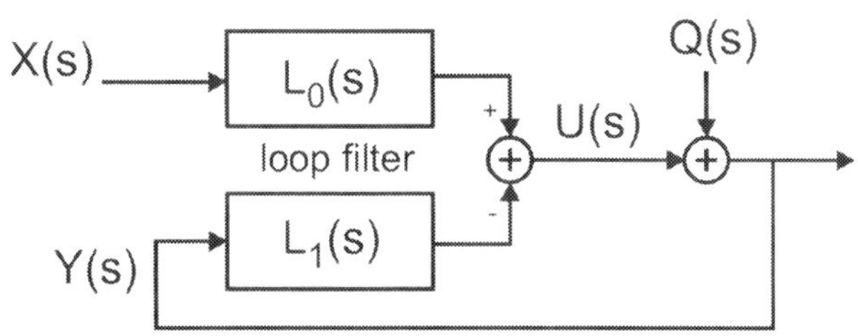

Figure 4-2: *Linear model of a ΣΔ modulator with a generic loop filter.*

$$NTF(s) = \frac{1}{1 - L_1(s)} \tag{4-1}$$

$$STF(s) = \frac{L_0(s)}{1 - L_1(s)} \tag{4-2}$$

The feedback loop filter $L_1(s)$ controls the trade-off between stability and noise shaping. For a fixed filter order N and sampling frequency f_S, the higher the low-frequency gain of $L_1(s)$, the greater the attenuation of quantization noise in the signal band. At the same time, the phase-shift in the modulator feedback path is increased. A stable N^{th}-order loop filter $L_1(s)$ has a -20N dB/dec roll-off at low frequencies, while at high frequencies it behaves like a 1st-order system with almost -90° phase-shift at $f_S/2$. If the feedforward loop filter $L_0(s)$ is made independent of $L_1(s)$, it is possible to control the *STF* frequency response to some extent. Different loop filter topologies for high-order frequency-compensated lowpass ΣΔ modulators are described in the following sub-sections.

4.1.1 Feedback Compensation

A high-order compensated loop filter can be achieved with multiple feedback architecture. This is depicted in Figure 4-3 for a single-bit 3rd-order CT ΣΔ modulator. In general, N feedback paths implement a loop filter $L_1(s)$ with N-1 zeros.

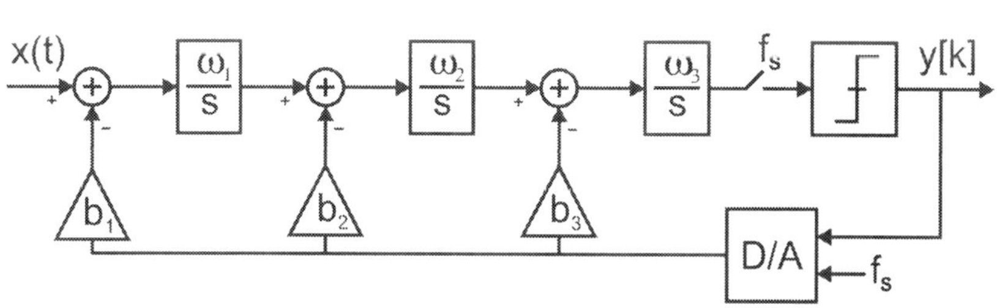

Figure 4-3: *Feedback compensated 3rd-order CT ΣΔ modulator.*

The loop filter transfer functions $L_0(s)$ and $L_1(s)$ are expressed by:

$$L_0(s) = \frac{\omega_1 \omega_2 \omega_3}{s^3} \tag{4-3}$$

$$L_1(s) = \frac{s^2 b_3 \omega_3 + s b_2 \omega_2 \omega_3 + b_1 \omega_1 \omega_2 \omega_3}{s^3} \tag{4-4}$$

where ω_i is the unity gain frequency of the ith integrator Int_i. The effect of high-frequency compensation on $L_1(s)$ can be seen in the Bode plots of Figure 4-4. This loop filter is stable when the sampling frequency f_S is such that the phase-shift at $f_S/2$ is close to -90°. The expressions for the *STF* and *NTF* are then derived from (4-3) and (4-4):

$$NTF(s) = \frac{s^3}{s^3 + s^2 b_3 \omega_3 + s b_2 \omega_2 \omega_3 + b_1 \omega_1 \omega_2 \omega_3} \tag{4-5}$$

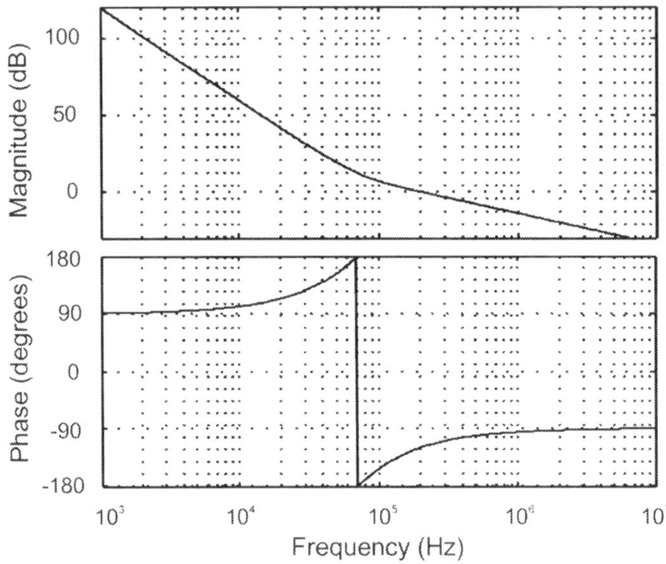

Figure 4-4: *Bode plots of the feedback compensated 3rd-order loop filter $L_1(s)$.*

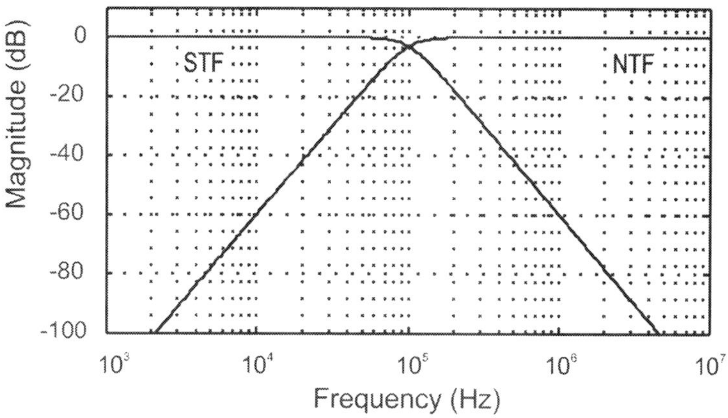

Figure 4-5: *STF and NTF frequency responses for the 3rd-order feedback compensated loop filter.*

$$STF(s) = \frac{\omega_1\omega_2\omega_3}{s^3 + s^2 b_3\omega_3 + s b_2\omega_2\omega_3 + b_1\omega_1\omega_2\omega_3} \tag{4-6}$$

The design of an Nth-order ΣΔ modulator normally starts with the selection of a suitable *NTF* based on the required quantization noise attenuation, the width (and location) of the signal band, and f_S. The *NTF* can be designed as a highpass Chebyshev or Butterworth filter. The coefficients b_i and ω_i are then determined on the basis of the loop filter stability and on the scaling of the ΣΔ modulator's internal voltages.

In this architecture, the *STF* is uniquely determined by the chosen *NTF* and by b_1, and it always presents a sharp N^{th}-order roll-off that can be used to reject high-frequency interferers. If b_1=1, the dc gain of the *STF* is 0dB. The *STF* and *NTF* are depicted for this case in Figure 4-5. Because the output bitstream is fed back to all loop filter nodes, the output of every integrator contains a strong replica of the input signal. Figure 4-6 shows the output spectrum of the three loop filter integrators for a -6dB sinusoidal input. As a consequence, the internal nodes of the ΣΔ modulator may experience relatively large voltage swings depending on the scaling of the filter coefficients.

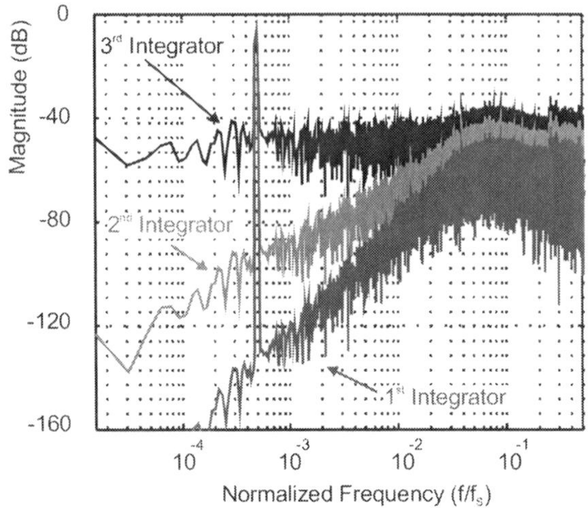

Figure 4-6: *Output spectrum of the 3 integrators of the feedback compensated loop filter.*

The purpose of scaling the loop filter coefficients is to keep the voltage swing of the internal nodes bounded to a certain limit (Figure 4-7). In this way, the integrators are prevented from operating beyond their designed linear region and, in so doing, keeping the ΣΔ modulator away from saturation. Due to the distributed feedback loop filter structure, the scaling of the coefficients imposes a relation $\omega_1<\omega_2<\omega_3$ among the integrators' unity gain frequencies for typical values of the b_i coefficients. Figure 4-8 shows the feedback compensated loop filter's input-referred transfer functions from each internal integrator. The noise from the 2nd integrator (Int_2) is magnified for the frequencies where the gain of the 1st integrator (Int_1) is below 0dB. In the same way, the noise of the 3rd integrator (Int_3) is also magnified for the frequencies where the combined gain of the preceding stages is below 0dB. Because of the scaling relation $\omega_1<\omega_2<\omega_3$, the value of ω_1 is often too small and the noise of the 2nd integrator is amplified inside the signal band. This is especially true in higher order modulators. In order to obtain a ΣΔ modulator whose thermal noise floor is dominated by the noise of the 1st integrator, the 2nd integrator also has to be designed for low noise and consequently higher power consumption. The feedback compensated architecture therefore enables the design of high-order stable ΣΔ modulators at the cost of extra power consumption.

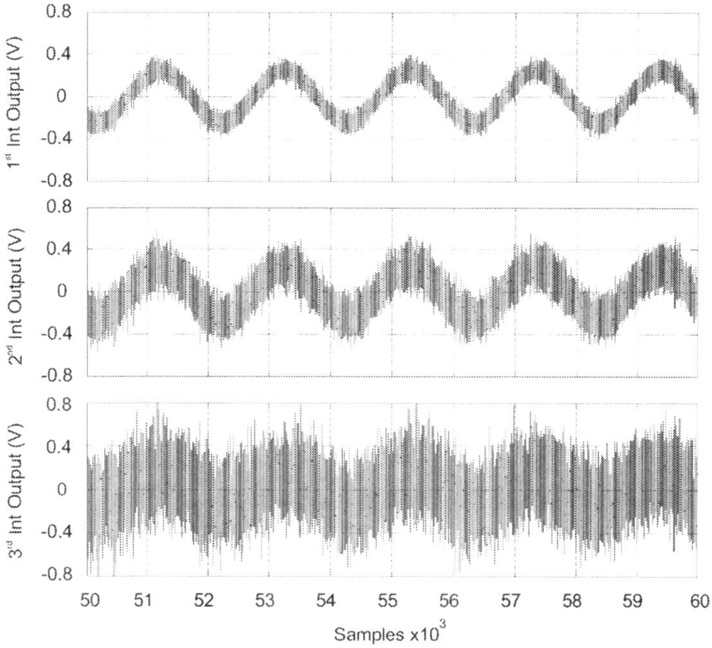

Figure 4-7: *Output voltage swing of the 3 integrators of the feedback compensated loop filter after scaling.*

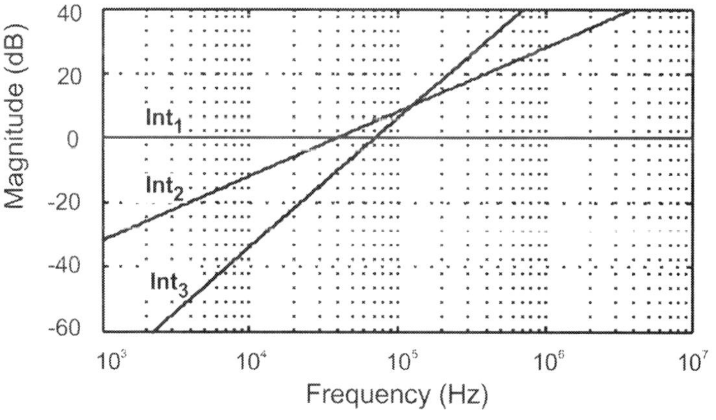

Figure 4-8: *Input-referred transfer functions of the 3rd-order feedback compensated loop filter.*

4.1.2 Feedforward Compensation

High-order stable loop filters can also be achieved by means of feedforward compensation (Figure 4-9). In this case there is only one feedback path and the loop filters $L_0(s)$ and $L_1(s)$ are the same:

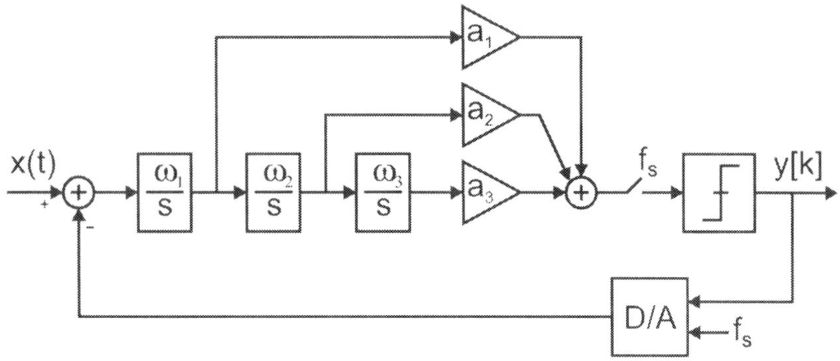

Figure 4-9: *Feedforward compensated 3rd-order CT ΣΔ modulator.*

$$L_0(s) = L_1(s) = \frac{s^2 a_1 \omega_1 + s a_2 \omega_2 \omega_1 + a_3 \omega_3 \omega_2 \omega_1}{s^3} \tag{4-7}$$

The resulting *NTF* and *STF* are given by:

$$NTF(s) = \frac{s^3}{s^3 + s^2 a_1 \omega_1 + s a_2 \omega_2 \omega_1 + a_3 \omega_3 \omega_2 \omega_1} \tag{4-8}$$

$$STF(s) = \frac{s^2 a_1 \omega_1 + s a_2 \omega_2 \omega_1 + a_3 \omega_3 \omega_2 \omega_1}{s^3 + s^2 a_1 \omega_1 + s a_2 \omega_2 \omega_1 + a_3 \omega_3 \omega_2 \omega_1} \tag{4-9}$$

Figure 4-10 shows the magnitude frequency response for the *NTF* and *STF*. Once again, after the *NTF* is chosen, the *STF* is completely defined. Major characteristics of the *STF* of the feedforward architecture are a 1st-order roll-off at high frequencies, and some peaking near the cut-off frequency. As a consequence, out-of-band interferers are poorly attenuated and even amplified around the peaking frequency.

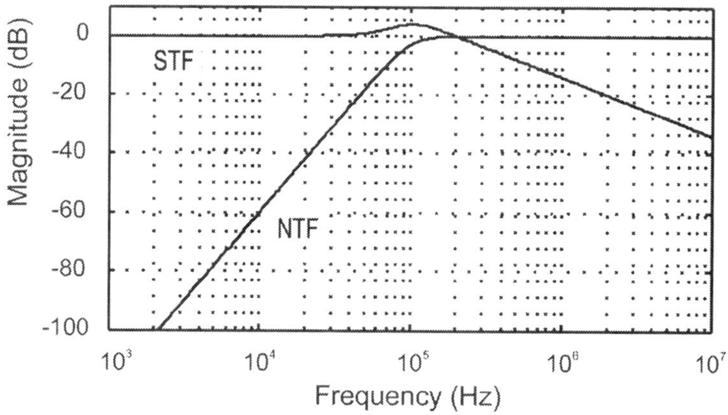

Figure 4-10: *STF and NTF frequency responses for the 3rd-order feedforward compensated loop filter.*

In this architecture, the loop filter only processes the error signal, and as a result, just an attenuated replica of the input signal is present at the various nodes of the loop filter. This fact becomes clear after inspecting the frequency spectrum of each integrator output (Figure 4-11). As a result, the internal nodes of a feedforward compensated loop filter are much less prone to large voltage swings than in the case of feedback compensation.

The mapping of a highpass Chebyshev or Butterworth filter to the *NTF* in (4-8) results in the definition of the coefficients a_i and ω_i. Again, the scaling of the coefficients defines the maximum excursion at every internal node. In contrast to the previous architecture, i.e. with a feedback compensated loop filter, the relation $\omega_1 > \omega_2 > \omega_3$ will generally apply to the integrators' unity gain frequencies. Because the power of the input signal is not strong at the output of the 1st integrator, ω_1 can be scaled to be larger than the ΣΔ modulator signal band. Consequently, the input-referred noise contribution of the high-order integrators are, in most cases, attenuated by the gain of the 1st integrator inside the band of interest. In a feedforward compensated loop filter, the input-referred transfer functions from each internal integrator are shown in Figure 4-12. As the ΣΔ modulator thermal noise floor is dominated by the 1st integrator, the high-order integrators can be designed with lower power consumption than in the case of feedback compensation.

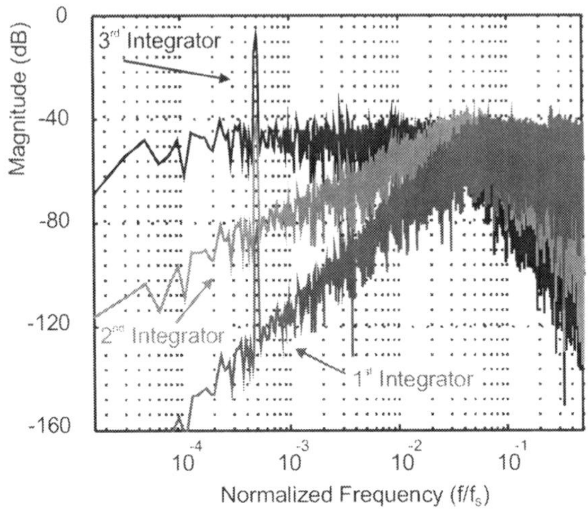

Figure 4-11: *Output spectrum of the 3 integrators of the feedforward compensated loop filter.*

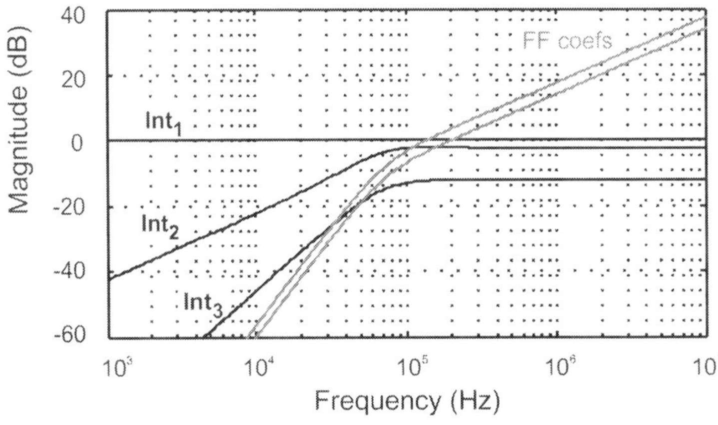

Figure 4-12: *Input-referred transfer functions of the 3rd-order feedforward compensated loop filter.*

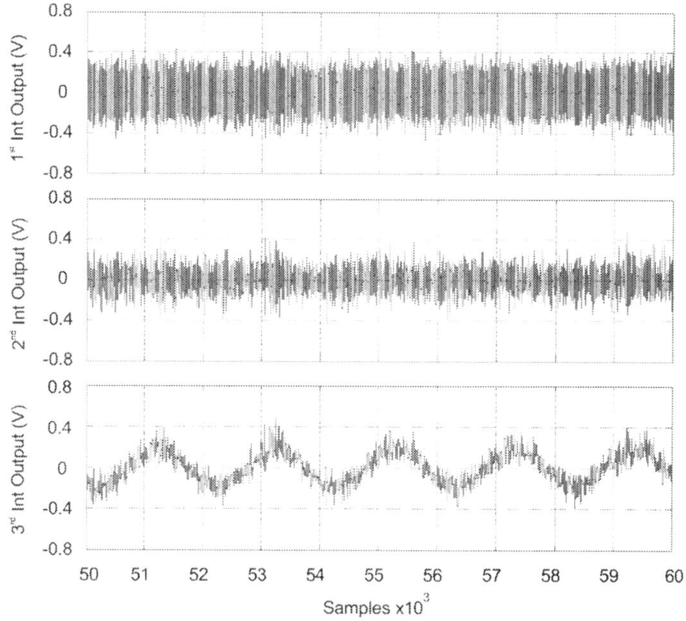

Figure 4-13: *Output voltage swing of the 3 integrators of the feedforward compensated loop filter after scaling.*

The internal voltages in a feedforward compensated ΣΔ modulator are shown in Figure 4-13 after scaling (ω_I=2π.200kHz). To achieve a similar result in a feedback compensated modulator, shown in Figure 4-7, it is necessary that ω_I=2π.40kHz.

4.1.3 Feedforward and Feedback Compensation

The feedforward and the feedback compensation techniques can be combined together (Figure 4-14) in order to obtain more control over the *STF*. The loop filter equations in this case are:

$$L_0(s) = \frac{s^2 a_1 \omega_1 + s a_2 \omega_2 \omega_1 + a_3 \omega_3 \omega_2 \omega_1}{s^3} \tag{4-10}$$

$$L_1(s) = \frac{s^2(a_1 b_1 \omega_1 + a_2 b_2 \omega_2 + a_3 b_3 \omega_3) + s(a_2 b_1 \omega_1 \omega_2 + a_3 b_2 \omega_2 \omega_3) + a_3 b_1 \omega_1 \omega_2 \omega_3}{s^3} \tag{4-11}$$

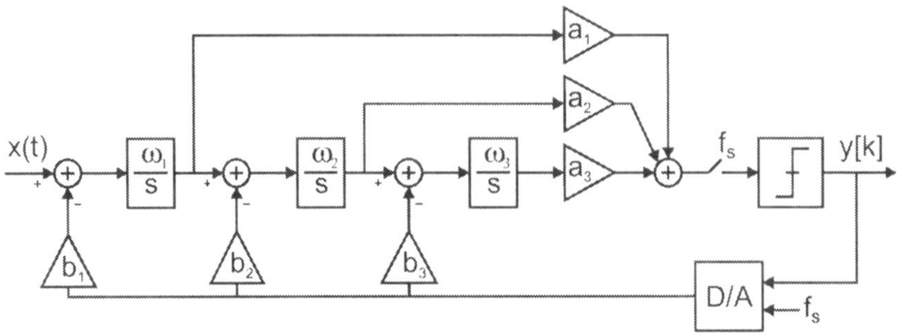

Figure 4-14: *Feedforward / feedback compensated 3rd-order CT ΣΔ modulator.*

And the resulting *NTF* and *STF* are given by:

$$NTF = \frac{s^3}{s^3 + s^2(a_1 b_1 \omega_1 + a_2 b_2 \omega_2 + a_3 b_3 \omega_3) + s(a_2 b_1 \omega_1 \omega_2 + a_3 b_2 \omega_2 \omega_3) + a_3 b_1 \omega_1 \omega_2 \omega_3}$$

(4-12)

$$STF = \frac{s^2 a_1 \omega_1 + s a_2 \omega_2 \omega_1 + a_3 \omega_3 \omega_2 \omega_1}{s^3 + s^2(a_1 b_1 \omega_1 + a_2 b_2 \omega_2 + a_3 b_3 \omega_3) + s(a_2 b_1 \omega_1 \omega_2 + a_3 b_2 \omega_2 \omega_3) + a_3 b_1 \omega_1 \omega_2 \omega_3}$$

(4-13)

The loop filter compensated with feedforward and feedback coefficients presents more degrees of freedom. As a result, its scaling is not as trivial as in the previous examples. For every unique set of coefficients a_i and b_i, four sets of integrator unity-gain frequencies $(\omega_1,\omega_2,\omega_3)$ can be found. A loop filter that combines the characteristics of the previous architectures can be designed. For example, the integrators' unity gain frequencies can be scaled for $\omega_1 > \omega_2$ and a *STF* with high-frequency roll-off greater than -20dB/dec can be obtained at the same time. This is possible because the *STF* numerator and the *NTF* can be designed together. Because there are more coefficients than the minimum required to map a Butterworth or Chebyshev highpass prototype to the *NTF*, the architecture can be simplified if needed. If some of the coefficients a_i or b_i are eliminated, the spectral content of the internal nodes change accordingly. Figure 4-15 shows two examples: for a_2=0 (black plot), a replica of the input signal with significant power is present at the output of all integrators and the scaling results in ω_I=2π.50kHz. For b_2=0 (gray plot), the output of the 1st integrator does not contain a strong replica of the input signal and ω_I=2π.137kHz can be obtained.

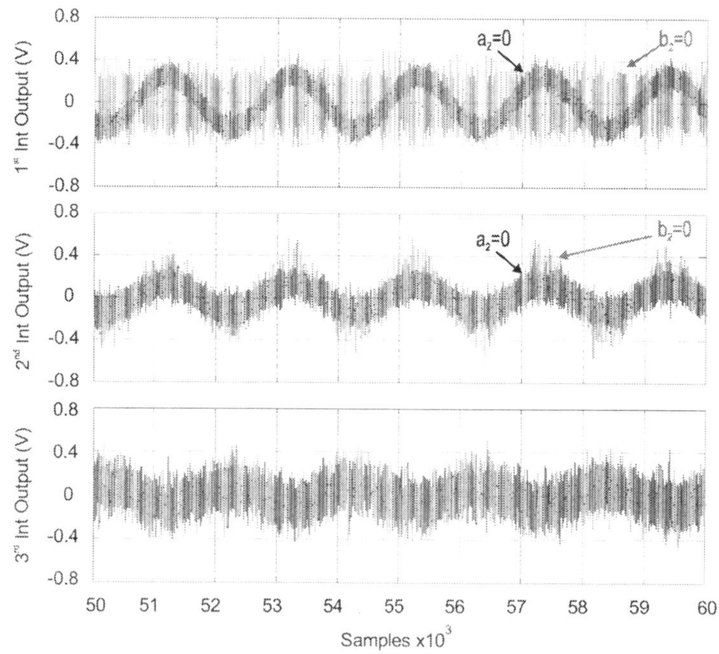

Figure 4-15: *Output voltage swing of the 3 integrators of the feedforward/feedback compensated loop filter for a_2=0 (black plot), and b_2=0 (gray plot) after scaling.*

Figure 4-16 shows three *STF* examples and the corresponding *NTF*. The loop filter with $a_1=0$ implements a *STF* with approximately $(N\text{-}1)^{th}$-order roll-off. The loop filter with $a_2=0$ implements a *STF* with very sharp roll-off and a notch at the cut-off frequency, but with just -20dB attenuation. In the case b_2=0, a 1st-order roll-off and some *STF* peaking are obtained, but the high-order integrators can be designed for lower power consumption.

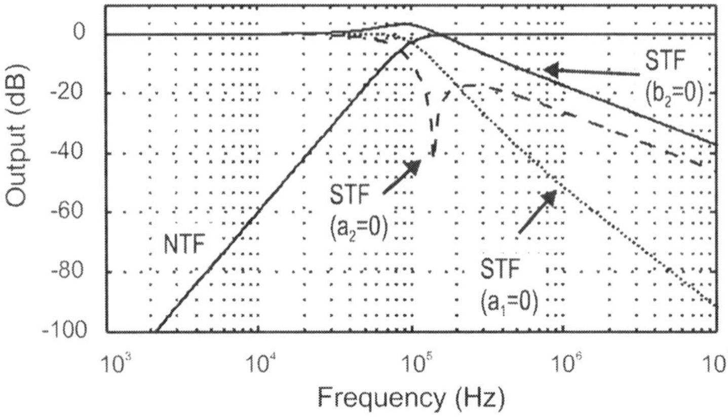

Figure 4-16: *STF and NTF frequency responses for the 3rd-order feedforward/feedback compensated loop filters.*

4.1.4 Feedback Compensation with Local Feedforward Path

Independent local feedforward paths can also be combined with a feedback compensated loop filter. The scaling of this architecture is simpler than the feedforward/feedback case as only one extra degree of freedom is added. The local feedforward path implements an extra zero in the *STF* and allows a loop filter scaling in which $\omega_1 > \omega_2$ if one of the b_i coefficients is eliminated from the architecture.

Figure 4-17 shows an example of a 3rd-order loop filter with a local feedforward path around the 2nd integrator (Int_2). The inclusion of the k_2 coefficient allows the elimination of the 2nd feedback path b_2, also enabling extra area and power consumption saving from a 2nd feedback DAC. The loop filter equations are:

$$L_0(s) = \frac{sk_2\omega_1\omega_3 + \omega_1\omega_2\omega_3}{s^3} \tag{4-14}$$

$$L_1(s) = \frac{s^2 b_3\omega_3 + s(b_2\omega_2\omega_3 + k_2 b_1\omega_1\omega_3) + b_1\omega_1\omega_2\omega_3}{s^3} \tag{4-15}$$

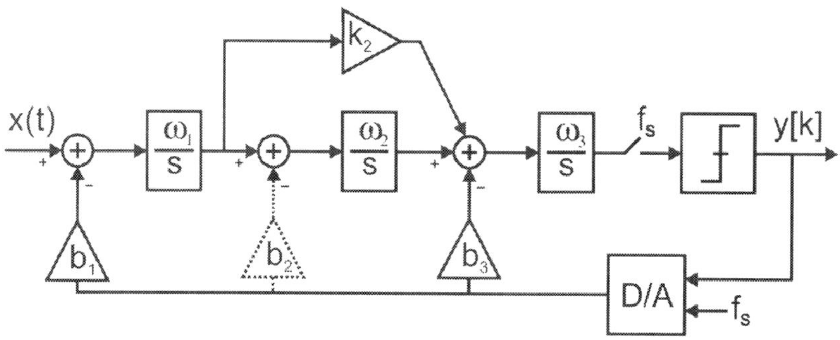

Figure 4-17: *Feedback compensated 3rd-order CT ΣΔ modulator with local feedforward path around Int_2.*

The resulting *NTF* and *STF* are given by:

$$NTF(s) = \frac{s^3}{s^3 + s^2 b_3 \omega_3 + s(b_2 \omega_2 \omega_3 + k_2 b_1 \omega_1 \omega_3) + b_1 \omega_1 \omega_2 \omega_3} \qquad (4\text{-}16)$$

$$STF(s) = \frac{s k_2 \omega_1 \omega_3 + \omega_1 \omega_2 \omega_3}{s^3 + s^2 b_3 \omega_3 + s(b_2 \omega_2 \omega_3 + k_2 b_1 \omega_1 \omega_3) + b_1 \omega_1 \omega_2 \omega_3} \qquad (4\text{-}17)$$

If b_2 is not removed from the loop filter, the possible range of values of k_2 (and the location of the *STF* zero s_z=-ω_2/k_2) is limited by the k_2<<($b_2\omega_2/b_1\omega_1$) constraint imposed by the location of the *NTF* poles. If b_2 is eliminated, a relation $\omega_3 > \omega_1 > \omega_2$ among the integrators' unity gain frequencies will be true in most cases. Figure 4-18 shows the input-referred transfer functions from each internal integrator for the feedback compensated loop filter ($b_2=0$) with local feedforward path. The frequency responses of the obtained transfer functions are somewhere between the frequency responses shown in Figure 4-8 and Figure 4-12.

Figure 4-19 shows two *STF* examples and the corresponding *NTF*. When all coefficients are present, a *STF* with monotonic (N-1)th-order roll-off is implemented. In the case $b_2=0$, a small amount of peaking can be observed near the cut-off frequency, but the same (N-1)th-order roll-off is still achieved. Local feedforward paths can be added around other integrators with similar results.

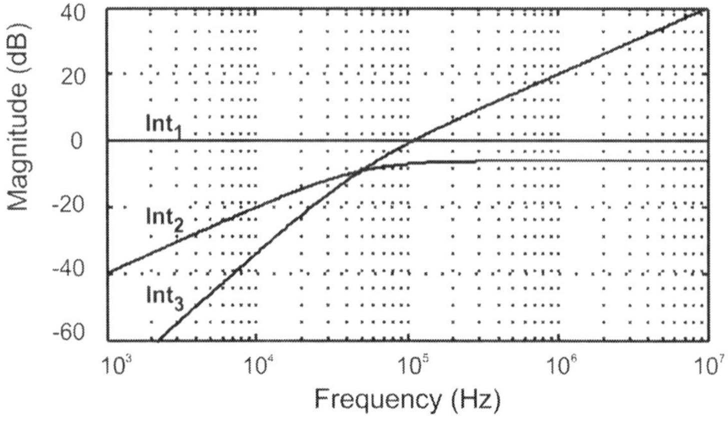

Figure 4-18: *Input-referred transfer functions of the 3rd-order feedback compensated loop filter with local feedforward coefficient and b_2=0.*

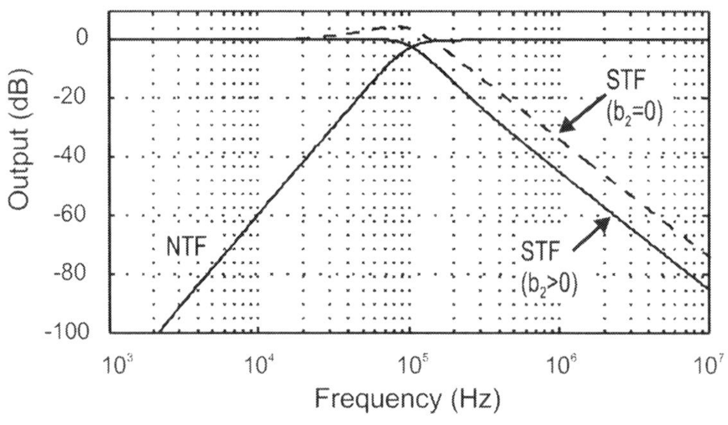

Figure 4-19: *STF and NTF frequency responses for the 3rd-order feedback compensated loop filter with local feedforward path around Int_2.*

4.1.5 Resonators and Local Feedback

A local feedback path around two loop filter integrators implements a resonator. In this way, part of the high dc gain of the loop filters of the previous architectures can be spread through the signal band. Figure 4-20 shows an example of a 4th-order feedforward compensated loop filter with two local feedback paths. The loop filter equation is:

$$L(s) = \frac{s^3 a_1 \omega_1 + s^2 a_2 \omega_2 \omega_1 + s(a_3 \omega_3 \omega_2 \omega_1 + d_2 a_1 \omega_4 \omega_3 \omega_1) + (a_4 + d_2 a_2)\omega_4 \omega_3 \omega_2 \omega_1}{(s^2 + d_1 \omega_1 \omega_2)(s^2 + d_2 \omega_3 \omega_4)} \tag{4-18}$$

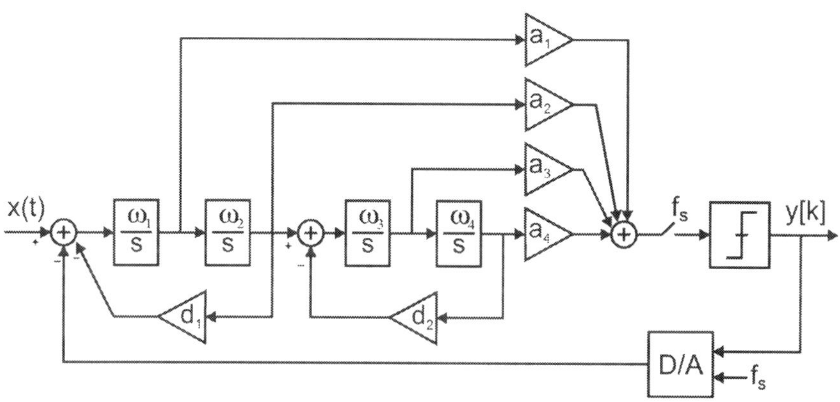

Figure 4-20: *Feedforward compensated 4th-order CT ΣΔ modulator with local feedback paths.*

The local feedback paths can be added to any other loop filter architecture with the same effect. As long as $d_2 << a_4/a_2$, the loop filter scaling remains unchanged. In this case the *NTF* and *STF* can be approximated by:

$$NTF(s) \cong \frac{(s^2 + d_1 \omega_1 \omega_2)(s^2 + d_2 \omega_3 \omega_4)}{s^4 + s^3 a_1 \omega_1 + s^2 a_2 \omega_2 \omega_1 + s a_3 \omega_3 \omega_2 \omega_1 + a_4 \omega_4 \omega_3 \omega_2 \omega_1} \tag{4-19}$$

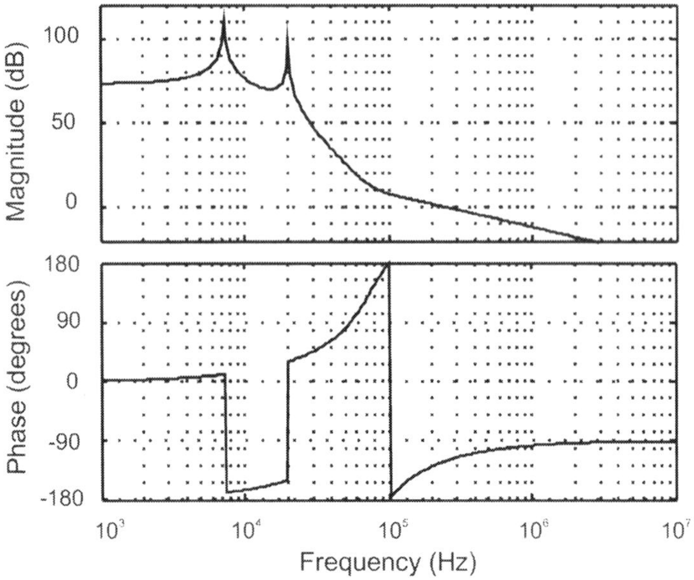

Figure 4-21: *Bode plots for the feedforward compensated 4th-order loop filter $L_1(s)$ with 2 resonators.*

$$STF(s) \cong \frac{s^3 a_1 \omega_1 + s^2 a_2 \omega_2 \omega_1 + s a_3 \omega_3 \omega_2 \omega_1 + a_4 \omega_4 \omega_3 \omega_2 \omega_1}{s^4 + s^3 a_1 \omega_1 + s^2 a_2 \omega_2 \omega_1 + s a_3 \omega_3 \omega_2 \omega_1 + a_4 \omega_4 \omega_3 \omega_2 \omega_1} \qquad (4\text{-}20)$$

The loop filter frequency response is plotted in Figure 4-21. Compared to a loop filter without resonators (Figure 4-4), the dc gain has been reduced. However two peaks corresponding to the resonant frequencies are present and the total quantization noise attenuation inside the signal band is increased. The *STF* and *NTF* frequency responses are plotted in Figure 4-22. The high-frequency phase response, and consequently the stability of the loop filter, remain the same. After the addition of the resonators, the modulator's stability can be verified through exhaustive transient simulations with large input signals at the resonant frequencies, which are expressed by:

$$\omega_{R12} = \sqrt{d_1 \omega_1 \omega_2}\ , \quad \omega_{R34} = \sqrt{d_2 \omega_3 \omega_4} \qquad (4\text{-}21)$$

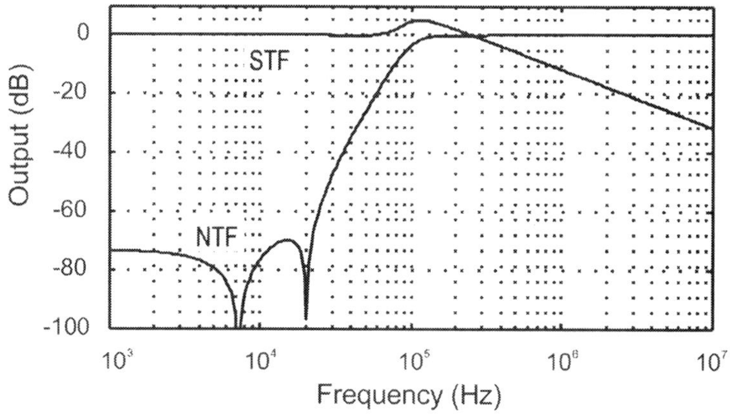

Figure 4-22: *STF and NTF frequency responses for the 4th-order feedforward compensated loop filter with two resonators.*

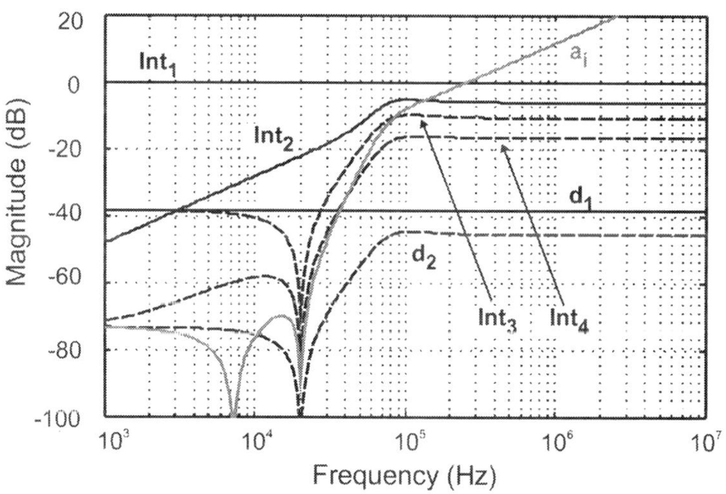

Figure 4-23: *Input-referred transfer functions of the 4th-order feedforward compensated loop filter with two resonators.*

The inclusion of the local feedback paths, normally implemented by transconductors in CT modulators, adds two extra thermal noise sources to the loop filter. The effect of these resonators in the ΣΔ modulator's input-referred transfer functions is depicted in Figure 4-23. The local feedback transconductors can be designed with low power consumption as their noise contributions are strongly attenuated at low frequencies. The noise contribution of the integrators remain unchanged at the edge of the signal band, where they are more significant.

4.2 IF-to-Baseband ΣΔ ADCs

To combine the high resolution and the relatively low power consumption of a lowpass ΣΔ modulator with IF digitization capabilities, the IF-to-baseband ΣΔ modulator was proposed [3]. This IF ADC consists of a mixer integrated with a lowpass ΣΔ modulator. The mixer can be placed inside [3], [5] or outside [4], [6] the ΣΔ loop.

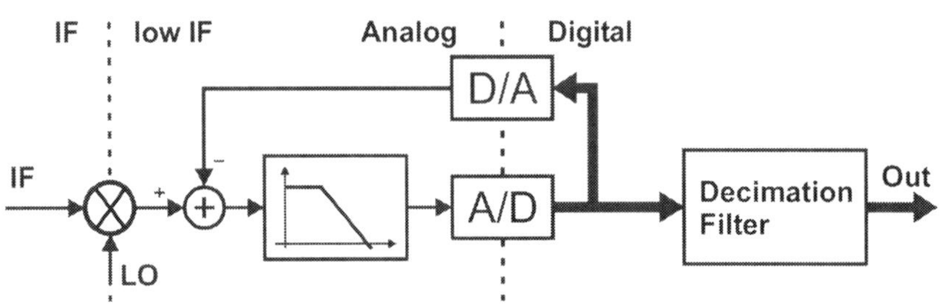

Figure 4-24: *IF-to-baseband ΣΔ ADC with mixer outside the loop filter for IF digitization.*

Figure 4-24 shows a CT lowpass ΣΔ modulator preceded by a mixer for frequency translation of the input signal. The analog input is down-converted from the IF to baseband and is then processed by the modulator. In principle, the mixer frequency (f_{LO}) and f_S can be chosen independently. As depicted in Figure 4-25, the wanted signal is also up-converted by the harmonics of the square wave LO driver. The CT loop filter acts as an anti-aliasing filter for high-frequency interferers. However, the mixing of undesirable signals at harmonics of the wanted signal with the harmonics of the LO signal cannot be avoided (Figure 4-26) and an attenuated replica of all the interferers will be down-converted to the baseband. Therefore, at least one stage of bandpass filtering is required between the antenna and the ADC. The mixer's conversion gain affects the maximum allowable input voltage of the complete system and should be taken into account in the design of the receiver's pre-amplification stages. The desired baseband frequency at the mixer output is given by:

$$f_{BB} = |f_{LO} - f_{IF}| \tag{4-22}$$

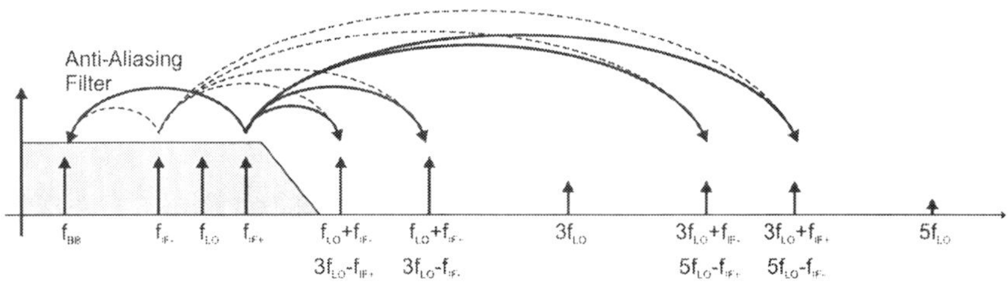

Figure 4-25: *CT loop filter acting as an anti-alias filter.*

Because the mixer is outside the ΣΔ modulator, its non-idealities are not attenuated by the loop gain. For example, the mixer linearity often determines the linearity of the whole system. Also, due to parasitic effects, the quantization noise power at frequencies near $f_S/2$ may leak to the mixer input causing an overall *SNR* degradation. This effect and other design issues regarding passive mixers for IF-to-baseband ADCs are discussed in detail in section 5.2.

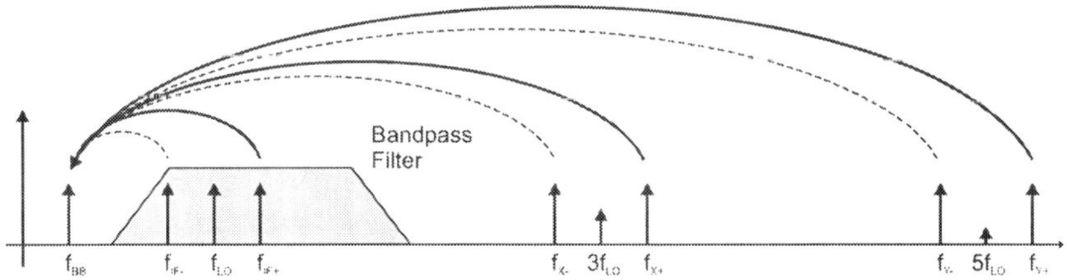

Figure 4-26: *Down-conversion of high-frequency signals by the harmonics of the mixer LO square wave driver.*

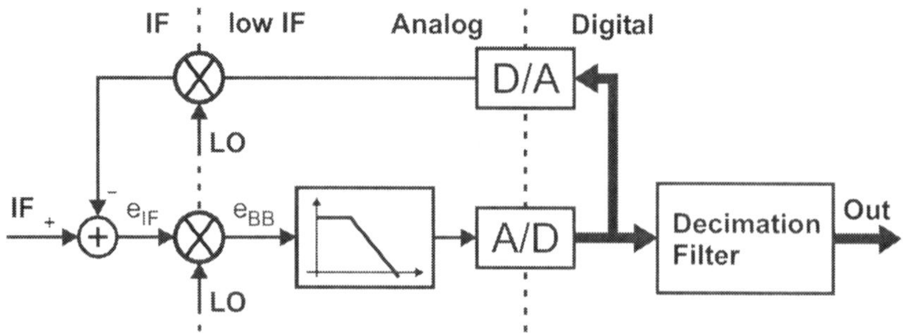

Figure 4-27: *IF-to-baseband ΣΔ modulator with mixer inside the loop filter for IF digitization.*

Figure 4-27 shows a CT ΣΔ modulator in which the mixer is placed inside the modulator loop. The analog input is again centred around a certain IF. Because the error signal e_{IF} is down-converted, another up-conversion mixer must be placed in the feedback path. After the frequency translation, the error signal e_{BB} is processed normally by the lowpass loop filter. The stability of the system is not disturbed because the effects of the feedforward and the feedback mixers cancel each other.

The main advantage of this topology is that non-idealities of the forward path mixer are attenuated by the modulator feedback loop. This is the case of ΣΔ modulators with multi-bit DAC or with a reconstruction filter in front of the single-bit DAC. As described in [5], suppression of mixer-induced distortion is not possible in ΣΔ modulators with standard single-bit feedback. In any case, the non-idealities of the feedback path mixer cannot be attenuated by the loop. However, the feedback path mixer can be implemented in the digital domain if f_{LO} is a rational fraction of f_S. The performance of the CT IF-to-baseband ΣΔ modulator in Figure 4-27 is sensitive to the delay between the mixers [7]. The quantization noise power around the LO frequency is down-converted to baseband during the delay interval, and the overall *SNR* is reduced. A detailed discussion about IF-to-baseband ADCs with mixer inside the loop filter can be found in references [5] and [7]. The use of IF-to-baseband ADCs in quadrature receivers is the subject of the next section. All examples discussed in section 4.3 are based on CT IF-to-baseband ΣΔ modulators with the mixer outside the loop filter.

4.3 Quadrature IF-to-Baseband ΣΔ ADCs

Figure 4-28 shows a detailed block diagram of a quadrature IF-to-baseband ADC, used in the low-IF quadrature receivers shown in Figures 2-6 and 2-7. This ADC is composed of two IF-to-baseband ADCs in parallel, with the mixers driven by quadrature signals LO_I and LO_Q. This ADC architecture combines the superior image rejection performance of quadrature-paths over single-path down-mixing (section 2.4), and the power efficiency and linearity of IF-to-baseband ΣΔ modulators [6].

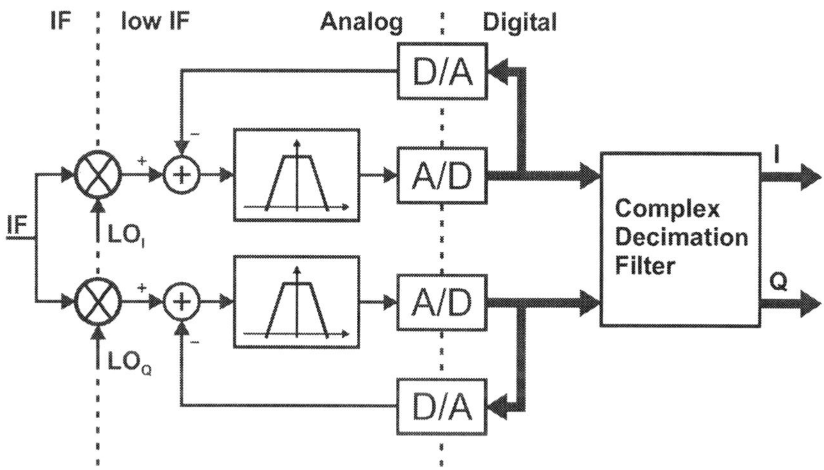

Figure 4-28: *Quadrature IF-to-baseband ΣΔ modulator with two real loop filters and analog mixers.*

Figure 4-29 shows the output spectrum (black plot) of a 4th-order quadrature CT IF-to-baseband ΣΔ modulator. Each loop filter is implemented with two real resonators, as depicted in Figure 4-20. The low-IF analog input obtained after quadrature mixing is a complex sinusoidal in the positive frequency band. The noise shaping, however, is still a symmetrical transfer function and the quantization noise is equally attenuated for positive and negative frequencies around dc. The *NTF* attenuation can be increased in one side of the signal band if a complex loop filter is used instead. Figure 4-30 shows a quadrature IF-to-baseband ΣΔ implemented with a complex loop filter.

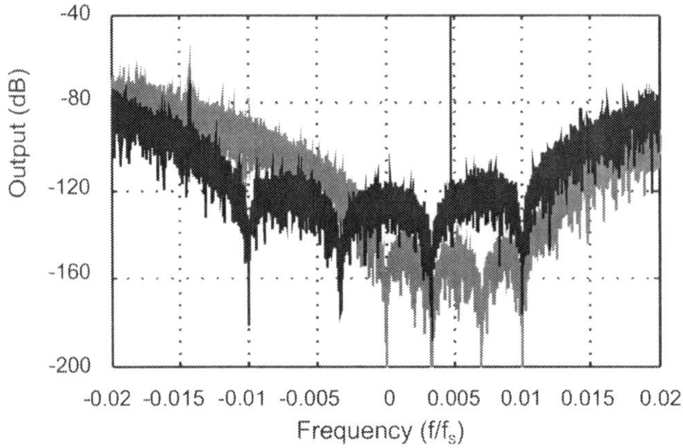

Figure 4-29: *Quadrature 4th-order ΣΔ modulator output spectrum with real noise shaping (black), and with complex noise shaping (gray).*

The gray plot in Figure 4-29 shows the output spectrum for the 4th-order quadrature CT IF-to-baseband ΣΔ modulator with a complex loop filter. The loop filter becomes an asymmetrical transfer function, and all *NTF* notches are placed in the positive frequency band.

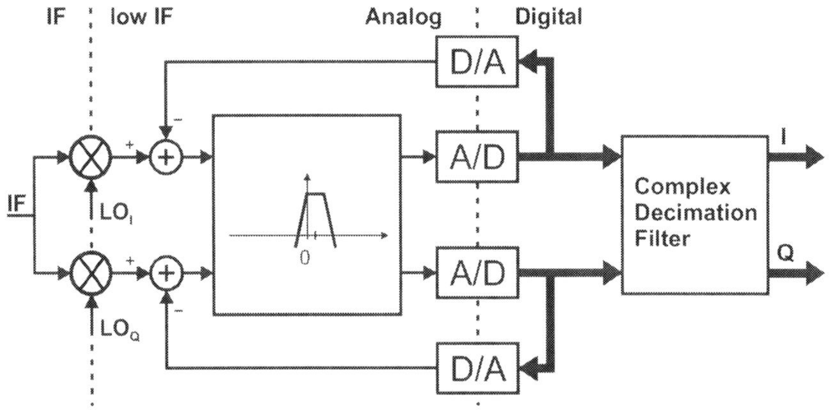

Figure 4-30: *Quadrature IF-to-baseband ΣΔ modulator with complex loop filter and analog mixers.*

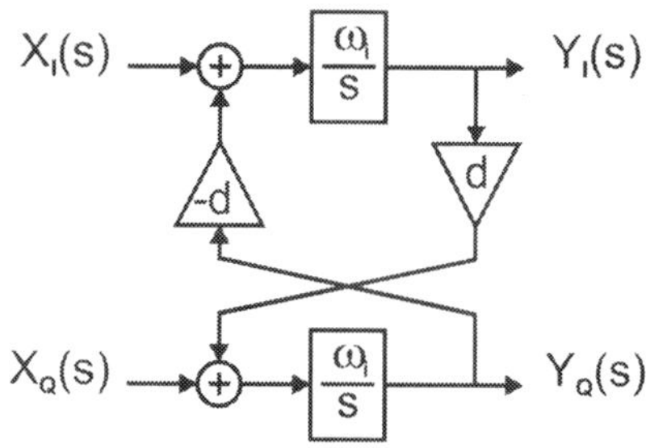

Figure 4-31: *Ideal complex resonator stage.*

The comparison between the output spectra shown in Figure 4-29 reveals the advantage of the complex noise shaping in quadrature ADCs. For the same modulator order, the quantization noise is 25dB more attenuated inside the positive side of the signal band. On the other hand, the modulator order can be reduced to achieve the same attenuation in the positive frequencies as with the real noise shaping in the complete signal band.

The main building block of a complex ΣΔ loop filter is the complex resonator [7]. For a complex input $X(s)$ and a complex output $Y(s)$, the complex resonator transfer function $H_C(s)$ is defined as:

$$H_C(s) = \frac{Y(s)}{X(s)} = \frac{Y_I(s) + jY_Q(s)}{X_I(s) + jX_Q(s)} = \frac{1}{\frac{s}{\omega_I} - jd} \tag{4-23}$$

where ω_I is the unity gain frequency of a real integrator and d is the complex resonator feedback coefficient. The transfer function (4-23) presents a single pole located at the positive resonant frequency ω_R and the unity gain frequency of the complex resonator is ω'_i. When $d \leq 0.3$, the shift on the unity gain frequency is less than 5% and there is no significant change on the loop filter phase response and stability. The relations among ω_I, ω_R, d and ω'_i are given by:

$$\omega_R = d \cdot \omega_I \tag{4-24}$$

$$\omega'_i = \omega_I \cdot \sqrt{1 + d^2} \tag{4-25}$$

The block diagram of an ideal complex resonator stage is shown in Figure 4-31. It is a system with two real integrators and cross-connection gain d. The equating of this two-dimensional linear system leads to:

$$\begin{bmatrix} Y_I \\ Y_Q \end{bmatrix} = \frac{1}{s^2 + d^2\omega_r^2} \cdot \begin{bmatrix} s\omega_r & -d\omega_r^2 \\ d\omega_r^2 & s\omega_r \end{bmatrix} \cdot \begin{bmatrix} X_I \\ X_Q \end{bmatrix} \tag{4-26}$$

If it is assumed that X_I, X_Q, Y_I and Y_Q are, respectively, the real and imaginary parts of the quadrature (complex) signals $X(s)$ and $Y(s)$, it can be shown that the linear system in (4-26) represents the complex resonator transfer function (4-23). Figure 4-32 shows the 4th-order complex loop filter used to generate the output spectrum shown in Figure 4-29 (gray plot). This quadrature ΣΔ modulator consists of two real 4th-order ΣΔ modulators connected by the cross-coupling coefficients d_1 to d_4. If one of the *NTF* notches is at dc, then one of the d_i coefficients can be nulled. In this case, d_1 is normally chosen to reduce power consumption as its thermal noise is the less attenuated from all cross-coupling coefficients.

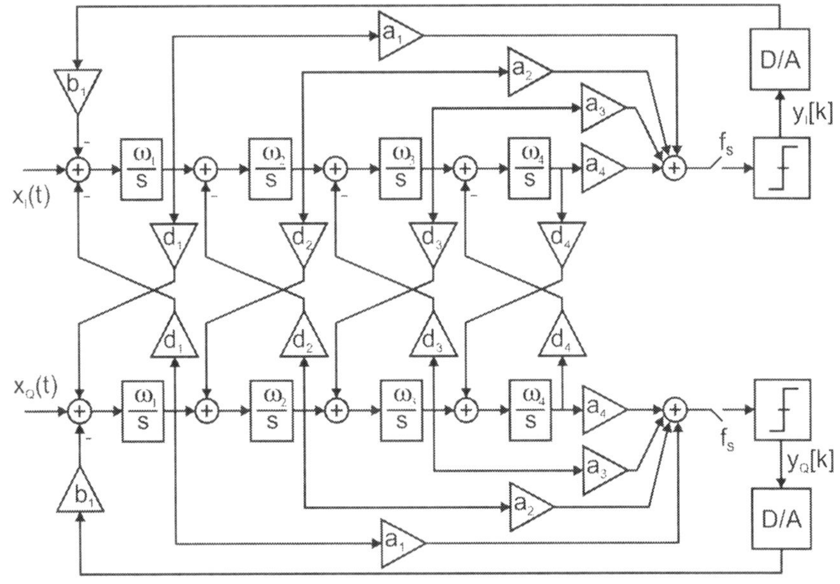

Figure 4-32: *Feedforward compensated CT 4th-order quadrature ΣΔ modulator with complex loop filter.*

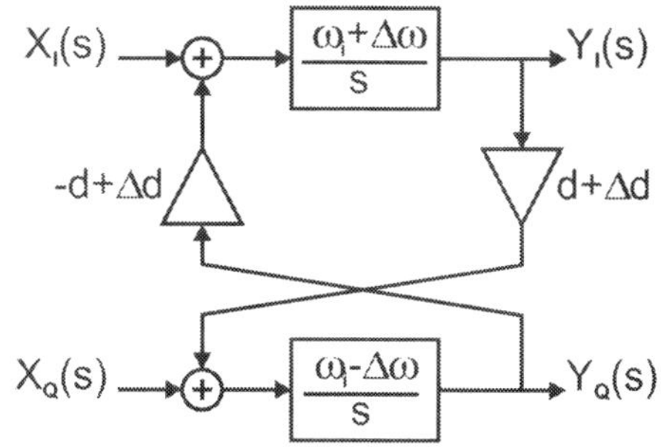

Figure 4-33: *Mismatched complex resonator stage.*

In analog filter implementations, however, it is not possible to achieve a perfect matching between the I and Q paths of a complex resonator. A more realistic model, accounting for integrator gain mismatch $\Delta\omega$ and cross-coupling gain mismatch Δd, is shown in Figure 4-33. A better insight on the effects of the I/Q imbalance is achieved through the equating of the mismatched complex resonator:

$$\begin{bmatrix} Y_I \\ Y_Q \end{bmatrix} = \frac{\begin{bmatrix} s\omega_I & -d\omega_I^2 \\ d\omega_I^2 & s\omega_I \end{bmatrix} \cdot \begin{bmatrix} X_I \\ X_Q \end{bmatrix} + \begin{bmatrix} s\Delta\omega & -\Delta d\omega_I^2 + d\Delta\omega^2 \\ \Delta d\omega_I^2 - d\Delta\omega^2 & s\Delta\omega \end{bmatrix} \cdot \begin{bmatrix} X_I \\ -X_Q \end{bmatrix}}{s^2 + (d^2 - \Delta d^2)(\omega_I^2 - \Delta\omega^2)} \tag{4-27}$$

The linear system in (4-27) reveals that the complex resonator output $Y(s)=Y_I(s)+jY_Q(s)$ can be expressed as a function of two complex inputs: $X(s)=X_I(s)+jX_Q(s)$ and its conjugate $X(s)^*=X_I(s)-jX_Q(s)$. According to the signal processing theory $X(f)^*=X(-f)$, i.e. the Fourier transform of complex conjugate signals are symmetrical related to dc [2]. This means some energy from the negative frequencies leak into the positive frequencies, and vice-versa. In the complex numbers notation, (4-27) can be rewritten as:

$$Y(s) = H_C(s) \cdot X(s) + H_{im}(s) \cdot X(s)^* \tag{4-28}$$

$$Y(s) = \frac{s\omega_I + jd\omega_I^2}{s^2 + (d^2 - \Delta d^2)(\omega_I^2 - \Delta\omega^2)} \cdot X(s) + \frac{s\Delta\omega + j(\Delta d\omega_I^2 - d\Delta\omega^2)}{s^2 + (d^2 - \Delta d^2)(\omega_I^2 - \Delta\omega^2)} \cdot X(s)^* \tag{4-29}$$

The resonator asymmetries lead again to an undesirable image problem. The image transfer function $H_{im}(s)$ defines the image power inside the signal band as a a function of the I/Q paths mismatch (Δd, $\Delta\omega$). The image rejection ratio (*IRR*) at the input of a complex resonator can be defined as:

$$IRR = \frac{H_{im}(s)}{H_C(s)} = \frac{s\Delta\omega + j(\Delta d\omega_1^2 - d\Delta\omega^2)}{s\omega_1 + jd\omega_1^2} \qquad (4\text{-}30)$$

If the relative implementation mismatch is defined as $\delta = \Delta d/d = \Delta\omega/\omega_1$ the *IRR* of a complex resonator can be calculated as a function of δ. Figure 4-34 shows the *IRR* at the input of a complex resonator as a function of δ. The effect of the 1st resonator mismatch on the output spectrum of a 4th-order quadrature modulator with 3 complex resonators can be seen in Figure 4-37a. The image band power is transferred to the signal band (and vice-versa) divided by the calculated *IRR* for different values of δ.

In the case of mismatch in the high-order resonators, the effect of the noise shaping should be considered for the estimation of the *IRR*. Figure 4-35 shows a complex linearized model of the 4th-order quadrature modulator shown in Figure 4-32, that takes in account complex resonator mismatch.

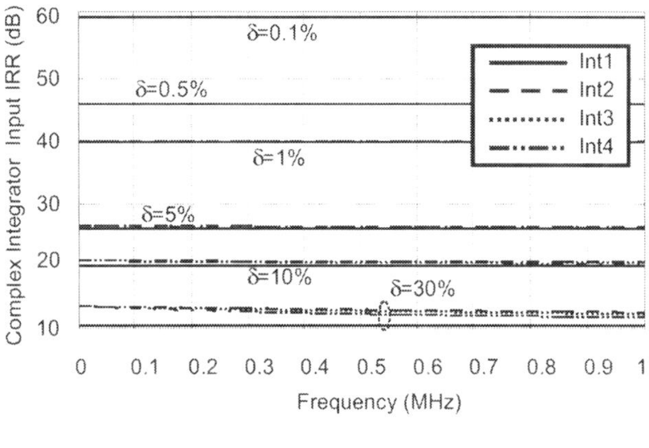

Figure 4-34: *Complex integrator IRR as a function of the relative mismatch* δ.

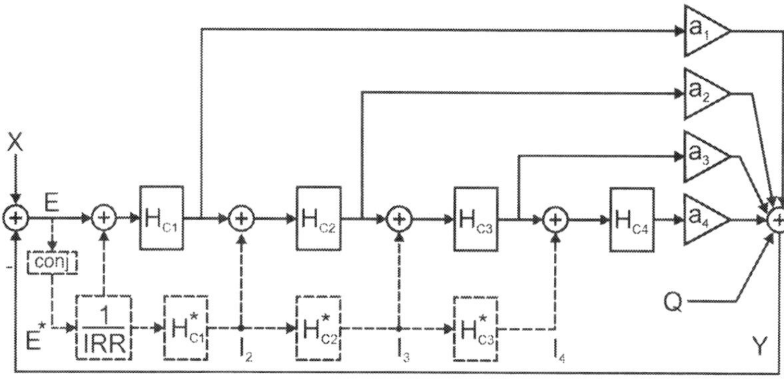

Figure 4-35: *Linear model of a 4th-order complex ΣΔ modulator with mismatched complex resonators.*

The transfer functions from the loop filter's resonators input to the ΣΔ modulator output $Y(s)/I_i(s)$ are depicted in Figure 4-36a. The complex conjugate input of each resonator is a replica of the complex error signal $E(s)$ divided by the *IRR* as defined in (4-30), and filtered by the conjugate transfer functions of the complex resonator $H_{Ci}(s)$. The effective *IRR* due to each mismatched resonator is obtained from the division of the values shown in Figure 4-34 by the transfer functions depicted in Figure 4-36b.

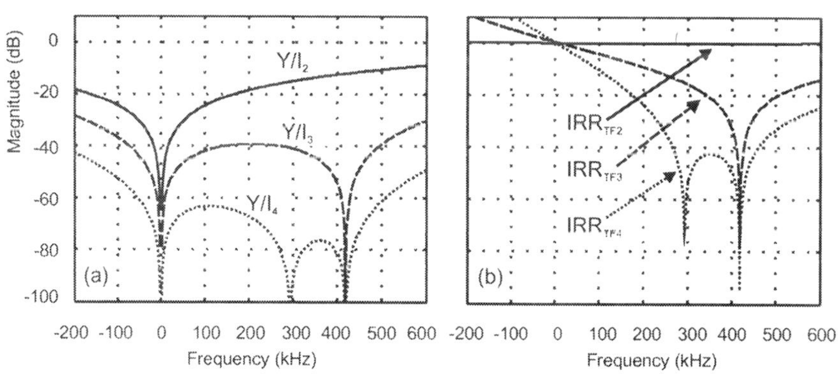

Figure 4-36: *Transfer functions from the input of each complex resonator to the ΣΔ modulator output (a), IRR transfer functions for each complex resonator (b).*

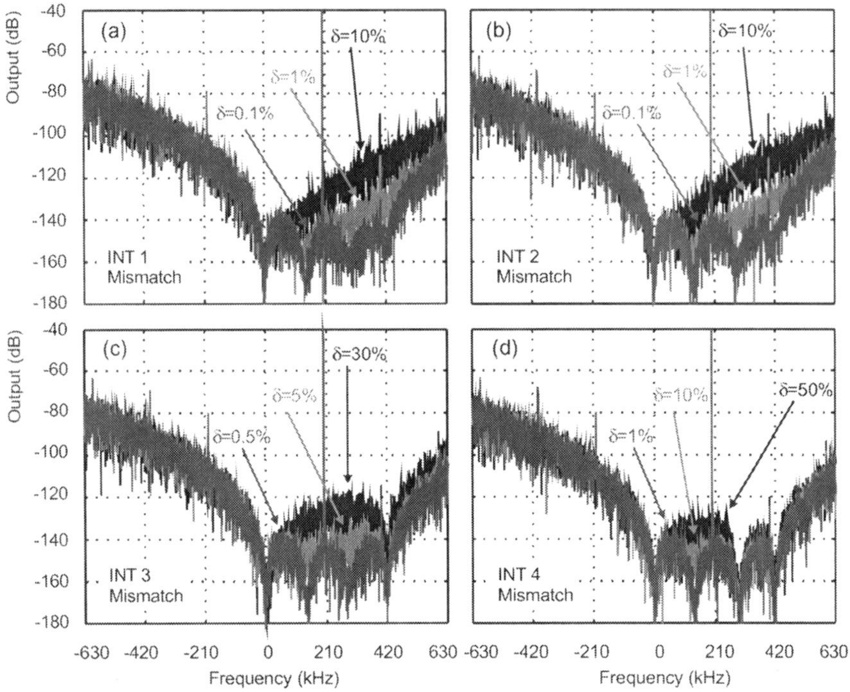

Figure 4-37: *Output spectrum of a 4th-order complex ΣΔ modulator as a function of the mismatch (δ) in each individual loop filter's complex resonator.*

Figure 4-37 shows the individual effect of the mismatch in each loop filter's resonator on the output spectrum of a 4th-order quadrature modulator. The simulated *IRR* for each resonator agrees with the value predicted by the analysis presented in previous paragraphs. It is important to notice that the input-referred *IRR* of the 2nd resonator is not weighted by the ΣΔ modulator noise shaping if the notch of the 1st resonator is at dc. The simultaneous effect of the mismatch in all loop filter's integrators in the modulator output is displayed in Figure 4-38 for different values of δ. The total noise power transferred from the image band to the signal band is the summation of the individual contributions from each resonator. The final amount of mismatch that can be tolerated in a quadrature ΣΔ modulator depends on the expected thermal noise floor.

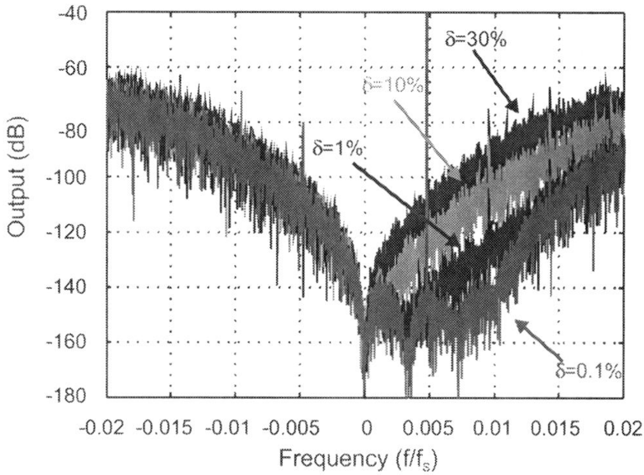

Figure 4-38: *Output spectrum of a 4th-order complex ΣΔ modulator as a function of the mismatch (δ) in all loop filter's complex integrators.*

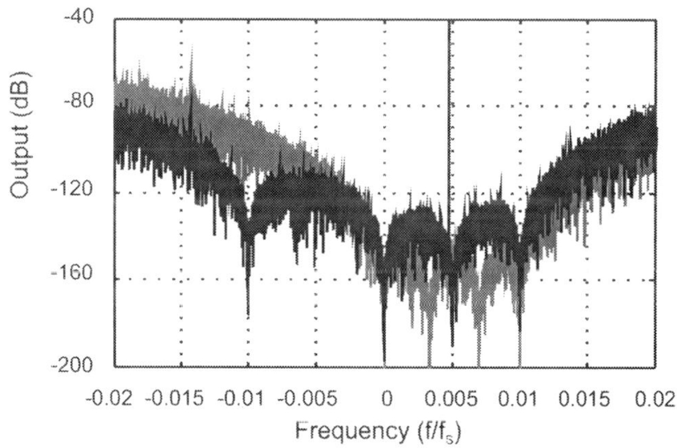

Figure 4-39: *Quadrature 4th-order ΣΔ modulator output spectrum with a single notch in the image band (black), and with full complex noise shaping (gray).*

A solution to increase immunity to the leakage of quantization noise into the signal band is the use of an image band notch (Figure 4-39, black plot). This strategy is simple: if it is not possible to further reduce the magnitude of the image transfer function $H_{im}(s)$ by design and prudent lay-out, the image notch can reduce the energy content of the image band. In this case the noise shaping in the signal band is reduced, but the attenuation in the image band makes the whole ΣΔ modulator less sensitive to the quadrature paths mismatch [23].

4.4 Bandpass ΣΔ ADCs

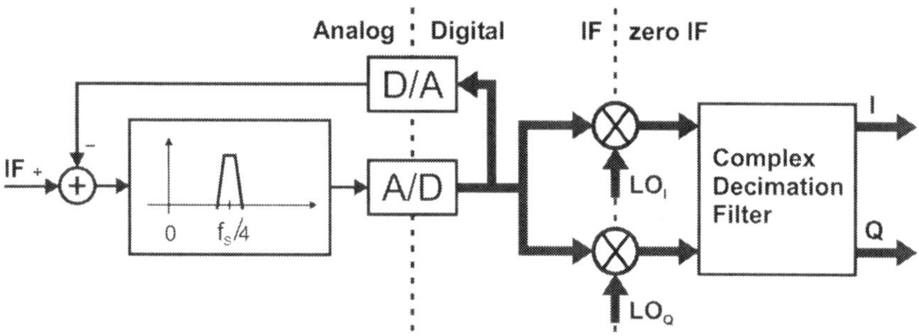

Figure 4-40: *Bandpass ΣΔ ADC for IF digitization with digital quadrature mixers.*

The first implementations of DSP based radios with A/D conversion at the IF level relied on bandpass ΣΔ ADCs [10]–[14]. As portrayed in the radio architecture of Figure 2-5, the receiver is a real (single-path) system until the bitstream generation and just one IF ADC is required. Figure 4-40 shows a detailed block diagram of bandpass ΣΔ ADC for IF digitization. The analog modulator is based on a real bandpass loop filter, while the zero IF down-conversion is implemented in the digital domain with almost perfect I/Q matching. The quadrature down-conversion relaxes the image suppression requirements of the channel selection filter, contributing to an overall cost reduction. The quadrature LO_I and LO_Q signals are then easily implemented in the digital domain in the case of zero-IF down-conversion ($f_{LO} = f_S/4$) [2]. Low-IF quadrature mixing can also be implemented in the digital domain at the cost of higher computational complexity.

4.4.1 Continuous-Time Bandpass ΣΔ Modulators

Bandpass ΣΔ modulators can be implemented by replacing the lowpass loop filter with a narrow bandpass filter. The center frequency (f_C) can be chosen arbitrarily inside the interval $0 < f_C < f_S/2$. It is usually a rational fraction of f_S, like $f_C = f_S/4$, in order to result in a simpler implementation. Because the signal band is far away from the low frequencies, bandpass modulators are immune to *1/f* noise and dc offset. For discrete-time systems, the transformation $z^{-1} \rightarrow z^{-2}$ generates a 2.Nth-order bandpass loop filter centred at $f_S/4$ from a Nth-order lowpass loop filter. Some high-order discrete-time bandpass ΣΔ modulators were described in the open literature [13]–[16]. An example is the 4th-order bandpass ΣΔ modulator generated from Candy's 2nd-order lowpass loop filter [2]:

$$L_{lp}(z) = \frac{2z^{-1} - z^{-2}}{(1 - z^{-1})^2} \rightarrow L_{bp}(z) = \frac{-2z^{-2} - z^{-4}}{(1 + z^{-2})^2} \qquad (4\text{-}31)$$

Continuous-time (CT) bandpass ΣΔ modulators are designed based on the *Impulse Invariance Transform* [17] applied to a discrete-time (DT) bandpass prototype [18]–[20]. This transformation generates a CT loop filter whose impulse response is equal to the impulse response of the DT prototype loop filter at the sampling instants. The resulting CT loop filter is also a function of the DAC pulse waveform: return-to-zero (RTZ), non return-to-zero (NRZ) or exponential from a switched-capacitor (SC) discharge. For example, the bandpass transfer function with $f_C = f_S/4$ in Eq. (4-31) is mapped to the following CT bandpass loop filter with NRZ DAC [18]:

$$L_{lp}(s) = \frac{1 + \frac{3}{2}T \cdot s}{T^2 s^2} \rightarrow L_{bp}(s) = \frac{\left(\frac{\pi}{2} - \frac{1}{4}\right)\frac{s^3}{T} + \left(\frac{3\pi^2}{16} + \frac{\pi}{4}\right)\frac{s^2}{T^2} + \left(\frac{\pi^2}{16} + \frac{\pi^3}{8}\right)\frac{s}{T^3} + \frac{3}{4}\left(\frac{\pi}{2T}\right)^4}{\left(s^2 + \left(\frac{\pi}{2T}\right)^2\right)^2} \qquad (4\text{-}32)$$

Figure 4-41 shows the magnitude frequency responses for $L_{lp}(s)$ and $L_{bp}(s)$ (f_S = 10MHz). The 2nd-order noise shaping around dc is replicated around f_C = ±2.5MHz. The effect of the lowpass-to-bandpass transformation on the continuous-time loop filter frequency response is clearly perceived.

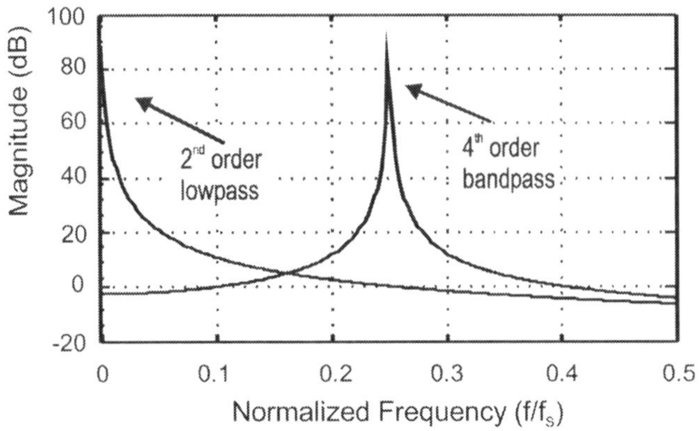

Figure 4-41: *Magnitude freq. response for $L_{lp}(s)$ 2nd-order lowpass and $L_{bp}(s)$ 4th-order bandpass loop filters.*

The architecture chosen to implement a CT bandpass modulator should allow the complete mapping of the transfer function $L_{bp}(s)$ to its coefficients. The feedback compensated chain of resonators (Figure 4-42) allows the implementation of the 4th-order $L_{bp}(s)$ as the feedback loop filter $L_1(s)$:

$$L_1(s) = \frac{s^3 b_4\omega_4 + s^2 b_3\omega_3\omega_4 + s(b_2\omega_2\omega_3\omega_4 + b_4 d_1\omega_1\omega_2\omega_4) + (b_1 + b_3 d_1)\omega_1\omega_2\omega_3\omega_4}{(s^2 + d_1\omega_1\omega_2)\cdot\ (s^2 + d_2\omega_3\omega_4)}$$

(4-33)

When the b_i and c_i coefficients are the same, the replicas of the input signal typically present at the output of every integrator in a feedback compensate loop filter are cancelled. As a result, the integrator's output spectra become similar to what is obtained with feedforward compensation (Figure 4-11). The input coefficients c_i also allow the control of the *STF* zeros independently of the *NTF*, as the feedforward loop filter $L_0(s)$ becomes:

$$L_0(s) = \frac{s^3 c_4\omega_4 + s^2 c_3\omega_3\omega_4 + s c_2\omega_2\omega_3\omega_4 + c_1\omega_1\omega_2\omega_3\omega_4}{(s^2 + d_1\omega_1\omega_2)\cdot\ (s^2 + d_2\omega_3\omega_4)} \qquad (4\text{-}34)$$

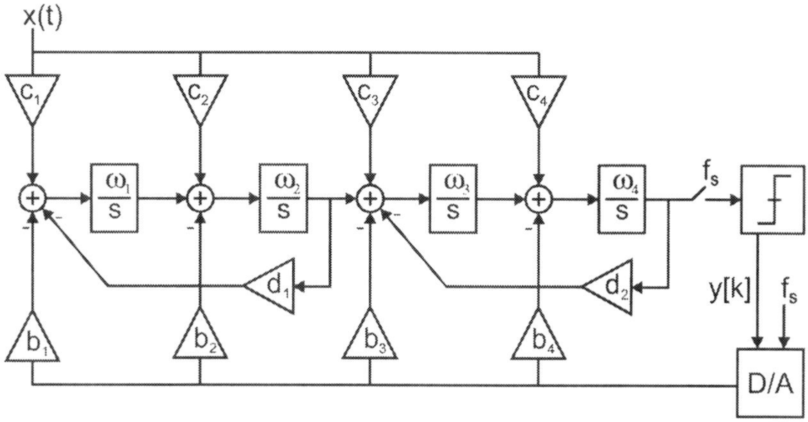

Figure 4-42: *Feedback compensated 4th-order CT ΣΔ modulator with multiple input coeff. and resonators.*

Figure 4-43 shows the resulting *NTF* (similar to Eq. (4-19)) and *STF* (c_3=c_4=0 in this example):

$$STF = \frac{s^3 c_4 \omega_4 + s^2 c_3 \omega_3 \omega_4 + s c_2 \omega_2 \omega_3 \omega_4 + c_1 \omega_1 \omega_2 \omega_3 \omega_4}{s^3 b_4 \omega_4 + s^2 b_3 \omega_3 \omega_4 + s(b_2 \omega_2 \omega_3 \omega_4 + b_4 d_1 \omega_1 \omega_2 \omega_4) + (b_1 + b_3 d_1)\omega_1 \omega_2 \omega_3 \omega_4} \tag{4-35}$$

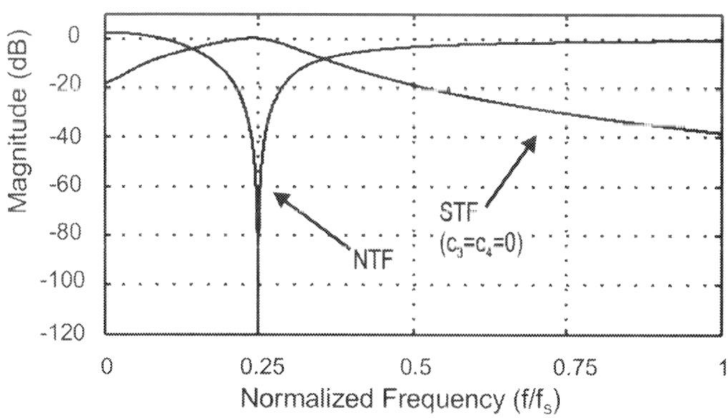

Figure 4-43: *STF and NTF for the 4th-order CT bandpass loop filter mapped to the mod. in Figure 4-42 (c_3=c_4=0).*

4.4.2 Continuous-Time Resonators

Resonators for both fully integrated lowpass and bandpass ΣΔ modulators are often implemented with a negative feedback path around two integrators [11]–[21]. An active-RC implementation is shown in Figure 4-44. A high input impedance G_m-C implementation is also possible at the expense of lower linearity [2]. The ideal resonator transfer functions ($A(s)=\infty$) are:

$$\frac{v_Y}{v_X} = \frac{\omega_I^2}{s^2 + \omega_I^2 \cdot R/R_R} \tag{4-36}$$

$$\frac{v_M}{v_X} = \frac{s\omega_I}{s^2 + \omega_I^2 \cdot R/R_R} \tag{4-37}$$

where $\omega_I = 1/RC$ is the integrator unity gain frequency and $d = R/R_R$ is the resonator feedback factor.

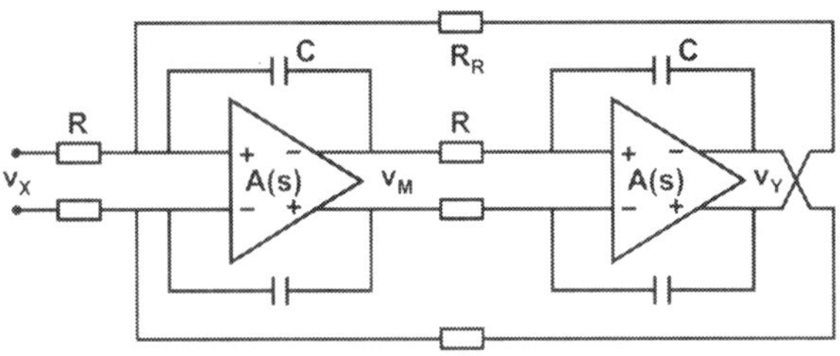

Figure 4-44: *Active-RC resonator implementation.*

The ideal resonator transfer functions present infinite quality factor. If the operational amplifier non-idealities are taken into account, a better insight is acquired about the resonator limits of performance. As in lowpass ΣΔ modulators, the quantization noise attenuation in the signal band is a function of the loop filter gain in the resonant frequency. Therefore the resonator quality factor limits the maximum achievable *SNR*. A more accurate opamp model is:

$$A(s) = \omega_u / s \tag{4-38}$$

where ω_u is the opamp unity gain radian frequency. When the opamp model of (4-38) is used, the resonator transfer functions become:

$$\frac{v_Y}{v_X} = \frac{\omega_i^2}{\frac{s^4}{\omega_u^2} + s^3\left(\frac{2\omega_i}{\omega_u^2} + \frac{2}{\omega_u}\right) + s^2\left(\frac{\omega_i^2}{\omega_u^2} + \frac{2\omega_i}{\omega_u} + 1\right) + d\omega_i^2} \qquad (4\text{-}39)$$

$$\frac{v_M}{v_X} = \frac{s^2 \cdot \omega_i / \omega_u^2 + s(\omega_i^2 / \omega_u^2 + \omega_i)}{\frac{s^4}{\omega_u^2} + s^3\left(\frac{2\omega_i}{\omega_u^2} + \frac{2}{\omega_u}\right) + s^2\left(\frac{\omega_i^2}{\omega_u^2} + \frac{2\omega_i}{\omega_u} + 1\right) + d\omega_i^2} \qquad (4\text{-}40)$$

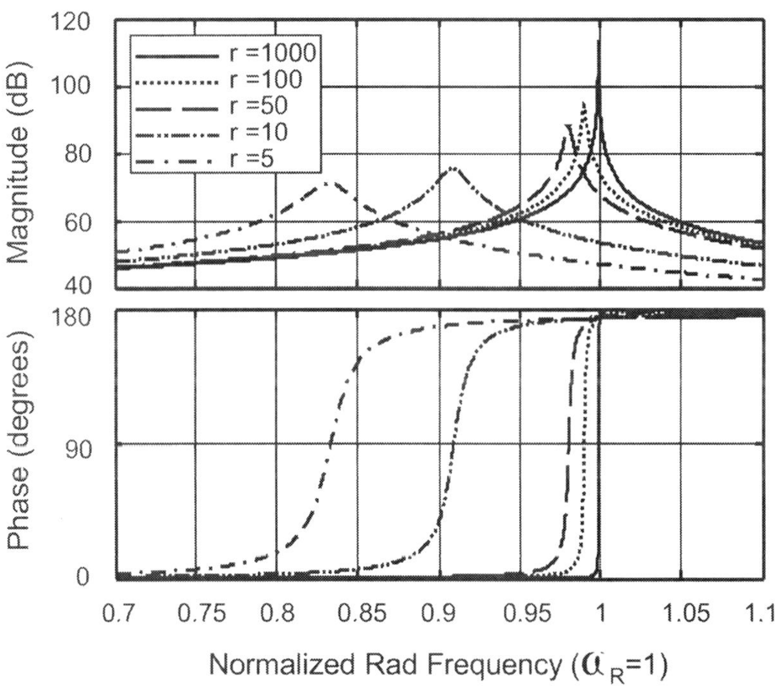

Figure 4-45: *Resonator normalized Bode plots as a function of* $r = \omega_i/\omega_u$. *The resonant frequency is* $\omega_R = 1$.

The ratio between the opamp's and the integrator's unity gain frequencies $r = \omega_i/\omega_u$ is a measure of how much the integrator phase-shift approaches the behaviour of a first or second order system. Figure 4-45 shows the normalised resonator Bode plots as a function of r. In this example some parameters were chosen: $\omega_R=1$, $d=0.01$ and $\omega_i=10$. If the opamp dc gain A_0 is finite, the resonator gain is always bounded by the opamp's dc gain. For large values of r, the relation $\omega_i << \omega_u$ is true and a very high quality factor is achieved. That is normally the case of low-frequency resonators used in lowpass ΣΔ modulators. For small values of r, the values for ω_i and ω_u are of the same order of magnitude. That is the case of high-frequency resonators used in bandpass ΣΔ modulators when the signal band is centred around a moderate to high IF. Then both the quality factor and the loop gain at the resonant frequency are limited. The deviation of the resonant frequency is also considerable.

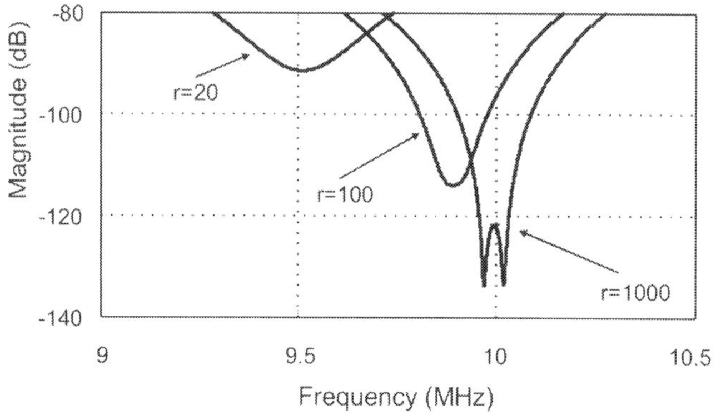

Figure 4-46: *Detail of the NTF of a 4th-order bandpass ΣΔ modulator with non-ideal opamps: r=20, r=100 and r=1,000.*

Figure 4-46 shows the calculated *NTF* of a 4th-order CT bandpass ΣΔ modulator implemented with non-ideal opamps (r=20, r=100 and r=1,000). The chosen center frequency of 10MHz is very close to the traditional FM radio IF (10.7MHz). In the case of almost ideal opamps (r=1,000), there is no degradation of the quantization noise attenuation and two *NTF* notches can be seen easily. However, when more realistic parameters are chosen (r=20: f_i=31.6MHz, f_u=632MHz) or (r=100: f_i=31.6MHz, f_u=3.16GHz), the *NTF* is clearly degraded. It is possible to compensate for the shift of the center frequency, but not for the reduced quantization noise attenuation. As a result, it is not possible to achieve the *SNR* predicted by calculations with ideal resonators described by (4-36) and (4-37).

When implemented with opamps that have the same gain-bandwidth (*GBW*) product and under the same *OSR*, signal bandwidth and loop filter topology, lowpass ΣΔ modulators can achieve higher resolutions than bandpass ΣΔ modulators. On the other hand, to digitize a certain bandwidth centred around a high IF with the same resolution of a lowpass modulator, a bandpass implementation generally consumes more power. This statement is better understood when the opamp unity gain frequency ω_u is expressed as a function of the amplifier transconductance (G_m) and load capacitance (C_L):

$$\omega_u = 2\pi f_u = G_m / C_L \tag{4-41}$$

For a given load capacitance, the only way to increase ω_u is by increasing G_m. Assuming the case of a single-stage CMOS opamp without source degeneration, the opamp G_m equals the transconductance g_m of each transistor in the differential input pair. Replacing G_m=$2I_D/(V_{GS}$-$V_T)$ in (4-41), the following is obtained:

$$\omega_u = \frac{2I_D}{(V_{GS} - V_T)C_L} \tag{4-42}$$

If the operation of the differential input pair in weak inversion is excluded, just modest increases in G_m can be obtained via reduction of V_{GS} in a low-voltage sub-μm CMOS technology. In this case, the increase of the bias current I_D, and consequently of the opamp power consumption, is the only way to achieve large increases in G_m and ω_u.

4.5 Quadrature Bandpass ΣΔ ADCs

The complex loop filter architecture shown in Figure 4-32 can also be used to build a quadrature bandpass ΣΔ modulator. In this case however, the complex resonator frequencies are placed far from dc. In a procedure similar to what is described in section 4.4.1, a complex bandpass *NTF* can be obtained from a complex lowpass *NTF* prototype.

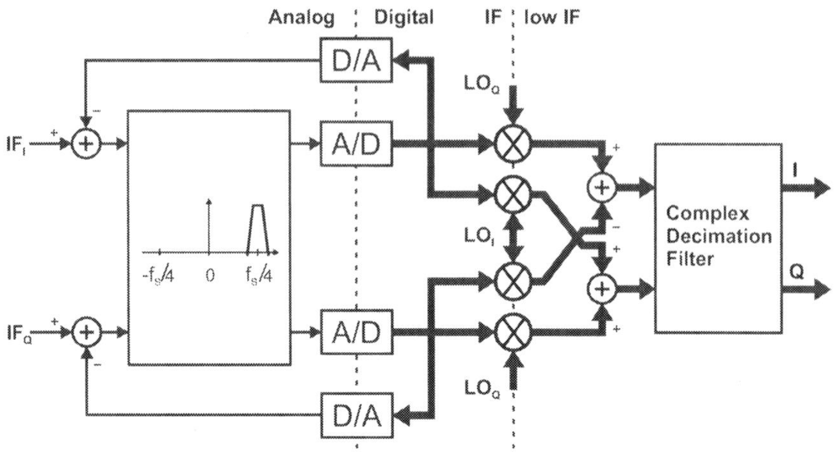

Figure 4-47: *Quadrature bandpass ΣΔ modulator with complex loop filter and full digital quadrature mixer.*

Figure 4-47 shows a detailed block diagram of a quadrature bandpass ΣΔ ADC. Such IF ADC architecture presents both quadrature inputs and outputs, and some implementations can be found in the open literature [23]-[25]. As it happens with real (single-path) bandpass ΣΔ modulators, down-mixing from the IF to a low-IF is implemented in the digital domain. Because the down-mixing operation in this cases involves the multiplication of two complex signals (IF_I+jIF_Q and LO_I+jLO_Q), four single-path mixers are necessary to implement a full digital quadrature mixer with almost perfect I/Q matching. However, an analog RF quadrature mixer is needed to generate the quadrature IF signal in the receiver front-end [25]. As a result, the digital I/Q matching advantage is lost as the front-end analog mixer presents finite image rejection.

As opposed to real (single-path) bandpass ΣΔ ADCs (Figure 4-40), quadrature bandpass ΣΔ ADCs are prone to leakage of quantization noise from the image band into the signal band. This happens because the analog bandpass loop filter is now a complex filter with finite I/Q paths matching. The mismatch induced I/Q cross-talk analysis presented in section 4.3 is also valid in this case. Again an image band *NTF* zero can be used to reduce leakage from the image band into the signal band [23].

4.6 Conclusions

This chapter presented four continuous-time ΣΔ ADC topologies that can be employed in DSP based radios for IF digitization: IF-to-baseband, quadrature IF-to-baseband, bandpass and quadrature bandpass. In this section, these four topologies are compared and the most suitable architecture for the AM/FM car radio application with reduced number of external components (Figure 2.29) is chosen.

Bandpass ΣΔ ADCs are capable of digitizing a narrow bandwidth directly at the IF. The digitized IF is then down-converted to a low-IF in the digital domain almost without errors (Figure 4-40). An IF-to-baseband ADC uses an integrated mixer to down-convert the desired channel to baseband. After the down-conversion, the desired channel is digitized by a lowpass ΣΔ modulator. As suggested by the analysis presented in section 4.4.2, bandpass ΣΔ modulators for IF digitization consume more power than lowpass ΣΔ modulators when fabrication technology, conversion bandwidth, loop filter architecture, *OSR* and the required dynamic range (DR) are the same. Table 4-1 compares several IF ADC designs with a 200 kHz BW centred around the 10.7 MHz IF. The assumption that IF-to-baseband ΣΔ ADCs are a more power efficient solution for IF digitization is enforced by the data in Table 4-1.

Table 4-1. *Comparison among several IF ADCs for FM radio.*

Ref.	DR (dB) in 200 kHz	IM3 (dB)	f_S (MHz)	Power (mW)	VDD (V)	Type
[13]	60	–	42.8	60	5.0	BP, SC
[15]	67	–	14.3	157	5.0	BP, SC
[19]	67	75	40	60	5.0	BP, CT
[26]	82	84	21.07	8	2.5	IF-to-B, CT
[16]	75	66	14.27	66	3.3	BP, SC
[21]	62	65	7.13	8.4	1.0	BP, SC
[20]	78	65	37.05	88	3.3	BP, SC
[27]	79	85	42.6	26	3.3	IF-to-B, CT

DSP based FM radio reception is a very common application for IF ADCs. The IF-to-baseband implementations [26], [27] achieve higher DR with less power consumption. The bandpass implementations with higher DR [20] consume 3 times more power than [27]. Furthermore, Table 4-1 also suggests that IF-to-baseband solutions are more linear. The reduced linearity makes bandpass ΣΔ ADCs unsuitable to be used for multi-channel AM reception. In this case, several AM channels are digitized at the same time. If the linearity is not high enough, the intermodulation products caused by very strong AM channels can be above the power level of a desired weak channel. As a result, the actual IF ADC dynamic range is reduced.

IF-to-baseband ADCs can be implemented with the mixer inside or outside the loop filter. A continuous-time loop filter implementation is preferred for this topology because of its intrinsic anti-alias filtering and lower power consumption compared to switched-capacitor implementations [1]. The linearity requirements (IM_3>85 dB) for multi-channel AM digitization demands the use of an intrinsically linear single-bit feedback DAC. At the same time, when the mixer is placed inside the loop filter, ΣΔ modulators with single-bit feedback do not compensate for the mixer-induced distortion [5]. These opposing requirements make the choice of placing the mixer inside the ΣΔ loop unattractive for the design presented in this book.

On the other hand, the use of a real (single-path) IF-to-baseband ADC requires an IF channel selection filter with very high image suppression. In order to relax the image suppression requirements of the IF channel selection filter, a quadrature IF-to-baseband ADC is preferred. A quadrature ΣΔ ADC requires two ΣΔ modulators to be implemented, doubling the effective power consumption. Even so, the quadrature IF-to-baseband solution is still more power efficient. As a matter of fact, the two IF-to-baseband solutions listed in Table 4-1 are quadrature ADCs and the nominal power consumption already accounts for the I and Q ΣΔ modulators [26], [27].

In quadrature IF-to-baseband ADCs, the quadrature low-IF down-mixing is performed in the analog domain. Therefore, the mismatch between the quadrature mixers result in finite image rejection for this architecture. In order to increase the quantization noise attenuation within the positive frequencies signal band, a complex loop filter is preferred (Figure 4-30). The required quantization noise attenuation could also be achieved with the two real loop filters solution (Figure 4-28) at the expense of increasing the loop filter order. This alternative would demand two 7th-order real ΣΔ modulators that require a higher *OSR,* and consequently internal amplifiers designed for a higher f_S. The mismatch between the I and Q paths inside a complex filter leads to the transfer of energy from the image band (negative frequencies) into the signal band (positive frequencies), and vice-versa. This undesirable effect decreases the *SNR*, but it can be greatly reduced if one of the *NTF* notches is placed at the edge of the image band. The AM/FM radio receiver requires a minimum *IRR* of 40 dB from the quadrature IF-to-baseband ADC. As demonstrated by [26]–[27], this can by achieved by careful lay-out.

Based on the above considerations, a CT quadrature IF-to-baseband ΣΔ ADC was chosen as the most suitable architecture for the car radio application with reduced number of external components. A 5th-order complex ΣΔ modulator with *OSR*=105 (f_S=42 MHz) and quadrature mixer placed outside the loop filter was chosen for this design. This choice enables the implementation of a power efficient ADC with enough linearity and DR. The system-level design and the circuit implementation of this IF-to-baseband ADC are presented and discussed in Chapter 5.

References

1. E.J. van der Zwan and E.C. Dijkmans, "A 0.2-mW CMOS ΣΔ Modulator for Speech Coding with 80-dB Dynamic Range", *IEEE J. Solid-State Circ.*, vol. 31, 1873–1880, Dec. 1996.
2. R. Schreier and G.C. Temes, "Understanding Delta-Sigma Data Converters", *IEEE Press, Piscataway, NJ*, 2005.
3. B-S. Song, "A Fourth-Order Bandpass Delta-Sigma Modulator with Reduced Number of Op Amps", *IEEE J. Solid-State Circ.*, vol. 30, 1309–1315, Dec. 1995.
4. F. Chen and B.H. Leung, "A 0.25-mW Low-Pass Passive Sigma-Delta Modulator with Built-In Mixer for a 10-MHz IF Input", *IEEE J. Solid-State Circ.*, vol. 32, 774–782, June 1997
5. A. Namdar and B.H. Leung, "A 400-MHz, 12-bit, 18-mW, IF Digitizer with Mixer Inside a Sigma-Delta Modulator Loop", *IEEE J. Solid-State Circ.*, vol. 34, 1765–1776, Dec. 1999.

6. L.J. Breems, E.J. van der Zwan, E.C. Dijkmans and J.H. Huijsing, "A 1.8mW CMOS ΣΔ Modulator with Integrated Mixer for A/D Conversion of IF Signal", *IEEE J. Solid-State Circ.*, vol. 35, 468–475, Apr. 2000.
7. L. Breems and J.H. Huijsing, "Continuous-Time Sigma-Delta Modulation for A/D Conversion in Radio Receivers", *Kluwer Academic, Boston, MA*, 2001.
8. L.J. Breems, E.C. Dijkmans and J.H. Huijsing, "A Quadrature Data-Dependent DEM Algorithm to Improve Image Rejection of a Complex ΣΔ Modulator", *IEEE J. Solid-State Circ.*, vol. 36, 1879–1886, Dec. 2001.
9. R. Schreier, "Quadrature Mismatch-Shaping", *Proceedings of the IEEE ISCAS*, May 2002, Volume 4, pages 675–678.
10. R. Schreier and M. Snelgrove, "Bandpass Sigma-Delta Modulation", *Electron. Lett.*, vol. 25, issue 23, 1560–1561, Nov. 1989.
11. S. Jantzi, R. Schreier and M. Snelgrove, "Bandpass Sigma-Delta Analog-to-Digital Conversion", *IEEE Trans. Circ. Syst.*, vol. 38, 1406–1409, Nov. 1991.
12. S. Jantzi, M. Snelgrove and P. Fergunson Jr., "A Fourth-Order Bandpass Sigma-Delta Modulator", *IEEE J. Solid-State Circ.*, vol. 28, 282–291, Mar. 1993.
13. F.W. Singor and M. Snelgrove, "Switched-Capacitor Bandpass Delta-Sigma Modulation at 10.7 MHz", *IEEE J. Solid-State Circ.*, vol. 30, 184–192, Mar. 1995.
14. A. Hairapetian, "An 81-MHz IF Receiver in CMOS", *IEEE J. Solid-State Circ.*, vol. 31, 1981–1986, Dec. 1996.
15. L. Louis, J. Abacarius and G.W. Roberts, "An Eighth-Order Bandpass ΔΣ Modulator for A/D Conversion in Digital Radio", *IEEE J. Solid-State Circ.*, vol. 34, 423–431, Apr. 1999.
16. P. Cusinato, D. Tonietto, F. Stefani and A. Baschirotto, "A 3.3-V 10.7-MHz Sixth-Order Bandpass ΣΔ Modulator with 74-dB Dynamic Range", *IEEE J. Solid-State Circ.*, vol. 36, 629–638, Apr. 2001.
17. J.A. Cherry and M. Snelgrove, "Continuous-Time Delta-Sigma Modulators for High-Speed A/D Conversion", *Kluwer Academic*, Boston, MA, 2000.
18. O. Shoaei and M. Snelgrove, "Design and Implementation of a Tunable 40 MHz-70 MHz Gm-C Bandpass ΔΣ Modulator", *IEEE Trans. Circ. Syst. II*, vol. 44, 521–530, July 1997.

19. J.A.E.P. van Engelen, R.J. van de Plassche, E. Stikvoort and A.G. Venes, "A Sixth-Order Continuous-Time Bandpass Sigma-Delta Modulator for Digital Radio IF", *IEEE J. Solid-State Circ.*, vol. 34, 423–431, Apr. 1999.
20. V. Colonna, G. Gandolfi, F. Stefani and A. Baschirotto, "A 10.7-MHz Self-Calibrated Swicthed-Capacitor-Based Multibit Second-Order Bandpass ΣΔ Modulator with On-Chip Switched Buffer", *IEEE J. Solid-State Circ.*, vol. 39, 1341–1346, Aug. 2004.
21. C-H. Kuo and S-I. Liu, "A 1-V 10.7MHz Fourth-Order Bandpass ΔΣ Modulators Using Two Switched Opamps", *IEEE J. Solid-State Circ.*, vol. 39, 2041–2045, Nov. 2004.
22. B. Razavi, "Design of Analog CMOS Integrated Circuits", *McGraw-Hill, New York*, 2001
23. S. Jantzi, K.W. Martin and A.S. Sedra, "Quadrature Bandpass ΔΣ Modulation for Digital Radio", *IEEE J. Solid-State Circ.*, vol. 32, 1935–1950, Dec. 1997.
24. F. Henkel et al., "A 1-MHz-Bandwidth Second-Order Continuous-Time Quadrature Bandpass Sigma-Delta Modulator for Low-IF Radio Receivers", *IEEE J. Solid-State Circ.*, vol. 37, 1628–1635, Dec. 2002.
25. R. Schreier et al., "A 375mW Quadrature Bandpass ΔΣ ADC with 90dB DR and 8.5MHz BW at 44MHz", *IEEE ISSCC Digest Technical Papers*, Feb. 2006.
26. E.J. van der Zwan, K. Philips and C.A.A. Bastiaansen, "A 10.7-MHz IF-to-baseband Sigma Delta A/D Conversion System for AM/FM Radio Receivers", *IEEE J. Solid-State Circ.*, vol. 35, 1810–1819, Dec. 2000.
27. Q. Sandifort, L.J. Breems, C. Dijkmans and H. Schuurmans, "IF-to-Digital Converter for FM/AM/IBOC Radio", Proceedings of the 29th European Solid-State Circuit Conference, Estoril, Portugal, Sept. 2003, pp. 707–710.

5 *IF-to-Baseband ΣΔ ADC for AM/FM/ IBOC Receivers*

The design of a single-bit 5th-order quadrature continuous-time IF-to-baseband ΣΔ ADC for AM/FM/IBOC car radios [1] is presented in this chapter. This ADC enables the realization of a quadrature low-IF receiver architecture with reduced number of external components described earlier in section 2.7.

Section 5.1 describes the quadrature IF-to-baseband conversion system together with the continuous-time ΣΔ modulator system-level design. The IF mixer design is the subject of section 5.2. Special attention is given to the issue of subsampling of the quantization noise. The design of the 1st integrator is presented in section 5.3, and the performance requirements for continuous-time and switched-capacitor integrators are discussed. Section 5.4 presents the most important design issues regarding the high-order integrators and resonators. The feedforward coefficients and the single-bit quantizer are described in section 5.5. Section 5.6 discusses the design of a single-bit switched-capacitor feedback DAC and the issue of jitter sensitivity in continuous-time ΣΔ modulators. This section also presents the design of a reference voltage generator for the SC feedback DAC. Experimental results from two fabricated IC prototypes are presented and analysed in section 5.7.

5.1 IF-to-Baseband Conversion System

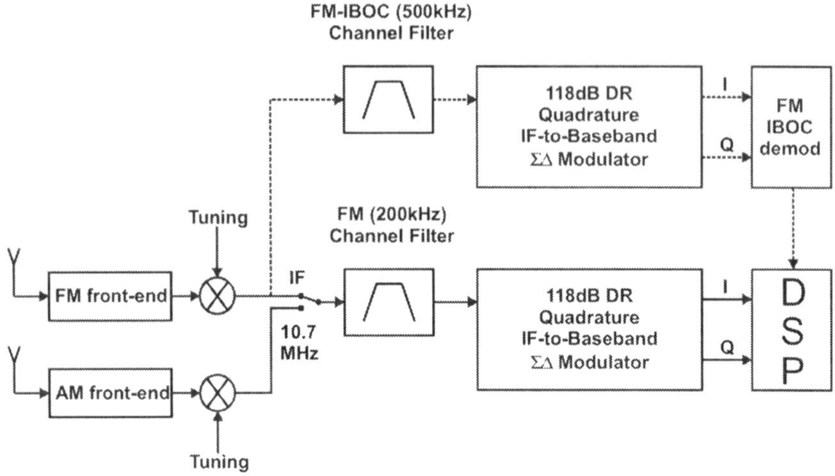

Figure 5-1: *AM/FM/IBOC car radio receiver with reduced number of external components.*

The proposed car radio receiver block diagram (Figure 5-1) is presented again in this chapter for the sake of completeness. Both AM and FM signals are processed by a 200kHz BW channel selection filter. Therefore, the IF-to-baseband ADC input consists of 1 FM channel or 20 AM channels. In order to receive the IBOC digital sidebands (section 2.6), a second IF-to-baseband ADC and an FM-IBOC digital demodulator have to be integrated together with the radio DSP [2]. Another ceramic channel selection filter (500 kHz BW) is also required for IBOC reception.

The IF-to-baseband conversion system with complex continuous-time ΣΔ modulators and integrated passive mixers for quadrature down-conversion is depicted in Figure 5-2. The 10.7MHz IF analog input signals are down-converted to a single-sided low-IF around +275kHz. The positive-frequencies signal band ranges from +175 to +375kHz in analog AM/FM mode, and from +25 to +525 kHz in FM-IBOC mode. The high-resolution complex baseband ADC runs at 41.7MHz. Two quadrature (I/Q) mixer driver signals at 10.425MHz are generated on-chip (divided by 4) from the ΣΔ modulator system clock [1].

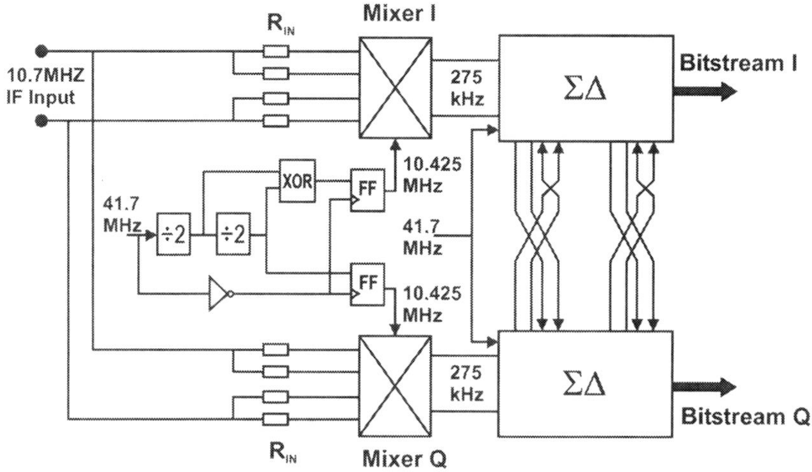

Figure 5-2: *IF-to-baseband conversion system.*

The receiver architecture shown in Figure 5-1 requires an IF ADC with 118dB DR in AM mode (3kHz BW), and 98dB DR in FM mode (200 kHz BW). The multi-channel AM digitization demands a very linear IF ADC with IM_3 > 85dB. For FM-IBOC reception, a SFDR > 70dB is required inside the 500 kHz signal band. The AM/FM receiver application demands 80dB of image rejection [3]. Because the FM channel filter has 40 dB out-of-band attenuation, an image rejection ratio (*IRR*) of at least 40 dB must be provided by the quadrature IF-to-baseband ΣΔ ADC. Table 5-1 summarizes the ADC design specifications.

Table 5-1. *IF-to-baseband ADC design specifications.*

	AM	FM	IBOC
Input IF	10.7MHz		
MAX Input	$0.5V_{RMS}$, fully-differential		
OSR	7000	105	42
BW	3kHz	200 kHz	500 kHz
DR	118dB	98dB	–
SFDR	–	–	>70dB
IM_3	>85dB		
IRR	>45dB		

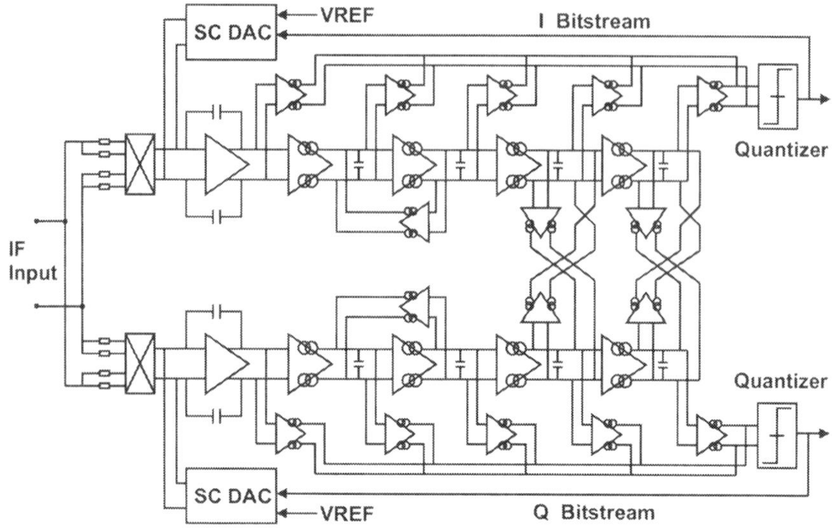

Figure 5-3: *Quadrature 5th-order 1b CT ΣΔ modulator.*

Figure 5-3 shows the single-bit 5th-order quadrature ΣΔ modulator topology used in this design. Two modulators are required to process the I and Q analog inputs. The quadrature mixers are placed between the input resistors and the virtual-ground nodes of the 1st OTA [4]. A continuous-time loop filter is chosen for this application because of its inherent anti-alias filtering characteristics. The high-frequency components generated by the mixers are then easily removed.

The 5th-order CT loop filter is stabilised by feedforward coefficients as they provide 1st-order phase-shift near half the sampling frequency (section 4.1.2). The graceful degradation of the loop filter guarantees large-signal stability through the clipping of the integrator outputs starting from the highest order integrator [5]. Real and complex resonators (Figure 5-3) provide the required quantization noise attenuation inside the single-sided positive-frequencies signal band. The real resonators are implemented through the local feedback path around the second and third integrators (section 4.1.5). The cross connections between the I and Q modulators implement two complex resonators (section 4.3). Figure 5-4 shows the loop filter Bode plots for positive and negative frequencies. The loop filter phase-shift is about 100deg at half the sampling frequency.

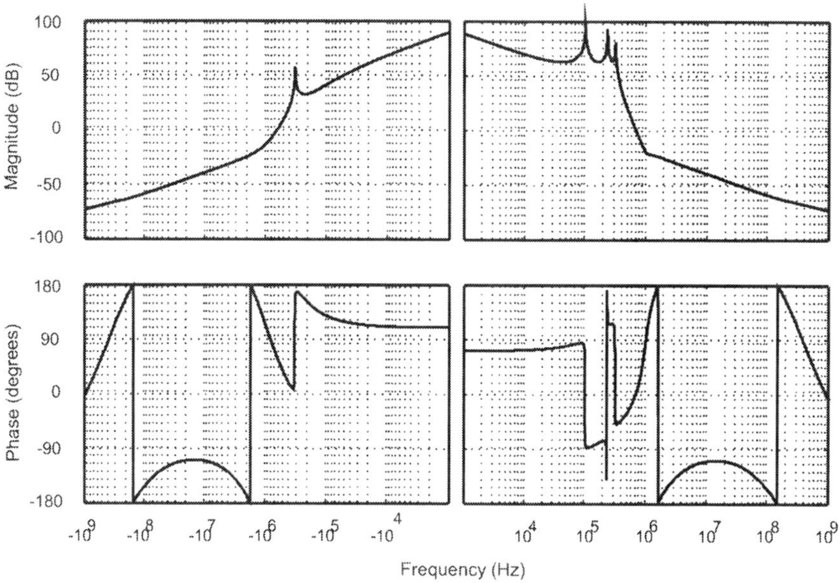

Figure 5-4: *Bode plots of the 5th-order CT complex loop filter (positive and negative frequencies).*

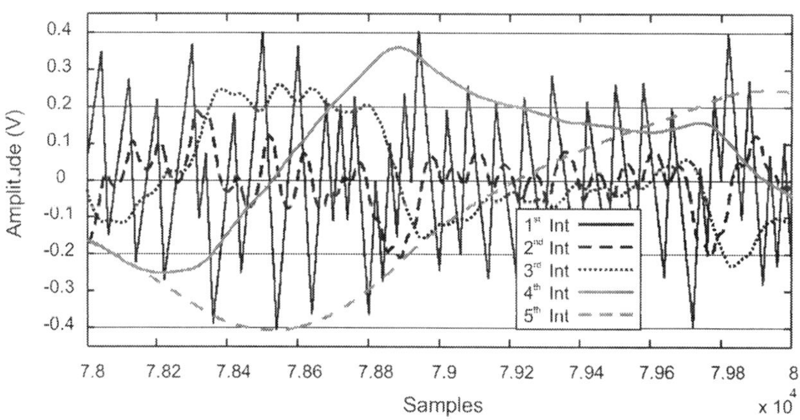

Figure 5-5: *Differential output voltages of the 5 loop filter's integrators for a 0.5 V_{RMS} ac input (simulation).*

The loop filter coefficients are scaled in order to guarantee that the differential output swing of each integrator is bounded to 0.4 V. This value was chosen to keep each loop filter OTA operating in the linear region, as the internal common-mode voltages were set to 0.7 V (V_{DD}=1.8 V). As a consequence, the 1st integrator unity gain frequency (ω_1) is 30% smaller than the 2nd integrator unity gain frequency (ω_2). Figure 5-5 shows the 5th-order ΣΔ loop filter internal voltages (simulation) for the maximum differential 0.5 V_{RMS} ac sinusoidal input. All internal states are bounded to the differential 0.4 V limit.

All the loop filter's building blocks contribute to the in-band noise budget and the quantification of dominant circuit noise sources is essential to achieve the required 118dB DR. Figure 5-6 shows the input-referred noise contribution of each loop filter building block. Because of the high gain of the 1st integrator at low frequencies, the circuit noise and the harmonic distortion from the other integrators are highly attenuated. The loop filter is designed to be thermal-noise-limited by the front-end circuitry. The input-referred noise of the 1st integrator and the DAC determine the ADC's noise floor at low frequencies. The input-referred noise of the 2nd integrator is also relevant at the edge of the signal bandwidth (175–375 kHz). This happens because, after the scaling of the coefficients, the gain of the 1st integrator is not high enough at these frequencies. The unity gain frequency of the 1st integrator could be increased if the sampling frequency is increased, however f_S=41.7 MHz is a system-level design constraint. The input-referred noise contribution from the high-order integrators, resonators and feedforward coefficients is designed to be marginal in the band of interest. The *1/f* noise (not shown in Figure 5-6) has a corner frequency of about 100 kHz. The maximum allowed thermal noise contribution determines the power consumption in each loop filter building block. For example, the 1st integrator is responsible for almost 70% of the total loop filter power consumption.

A single-bit quantizer is used together with a single-bit, inherently linear, switched-capacitor (SC) feedback DAC. In this case most of the feedback charge is integrated at beginning of the DAC period, therefore a continuous-time ΣΔ modulator implemented with a SC DAC can tolerate a larger delay between the quantizer and clock signals and is less sensitive to jitter-induced errors. This solution was proposed by [6] and successfully implemented in [7] for a radio receiver application.

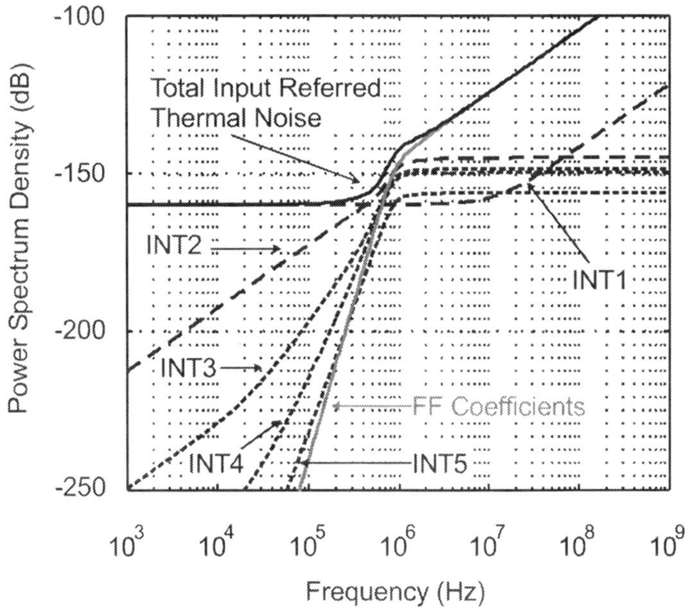

Figure 5-6: *Input-referred thermal noise contribution from each loop filter building block.*

The quadrature ΣΔ modulator' s simulated output spectrum is shown in Figure 5-7. The quantization noise in the right side of the spectrum is much lower than on the left side due to the three notches placed across the positive-frequencies single-sided signal band. Because the analog input is also a complex signal, the frequency location of the input tone is asymmetrical in relation to dc. The notch placed on the left side of the spectrum is very important because it reduces quantization noise leakage from the image band into the signal band due, for instance, to mismatch between the I and Q ΣΔ modulators (section 4.3).

The spread in the resonator's passive components (i.e., resistors and capacitors) cause deviation in the position of the notches. As a result, the quantization noise power increases inside the signal band. The impact of these non-idealities is small since, by design, the overall noise in the band of interest is dominated by circuit thermal noise (gray plot, Figure 5-7). In this way, even with maximum spread, the quantization noise will still be below the circuit noise.

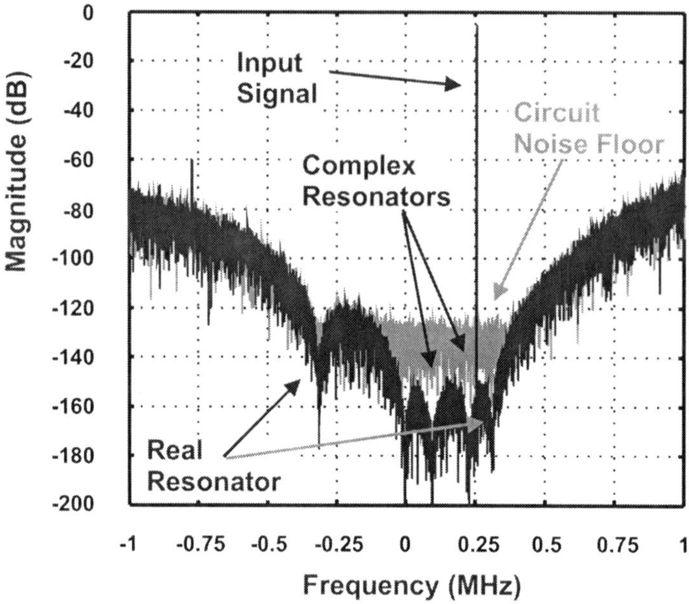

Figure 5-7: *Simulated output spectrum of the single-bit 5th-order CT quadrature ΣΔ modulator (RBW=200Hz).*

The following sections describe the most important circuit building blocks of this quadrature CT IF-to-baseband ΣΔ ADC: IF mixers, 1st integrator, high-order integrators and resonators, feedforward coefficients and single-bit quantizer, and switched-capacitor DAC. This chapter ends with the presentation and discussion of experimental results obtained from two fabricated IC prototypes [1], [8].

5.2 IF Mixer

Figure 5-8 shows the mixer topology used in this design to implement the low-IF quadrature down-conversion. An NMOS passive mixer is implemented together with an active RC integrator. The mixer switches are placed between the input resistors and the virtual-ground nodes of the OTA [4]. This combined mixer/integrator implementation is used as the input stage for an IF-to-baseband CT ΣΔ modulator with the mixer outside the loop (section 4.2).

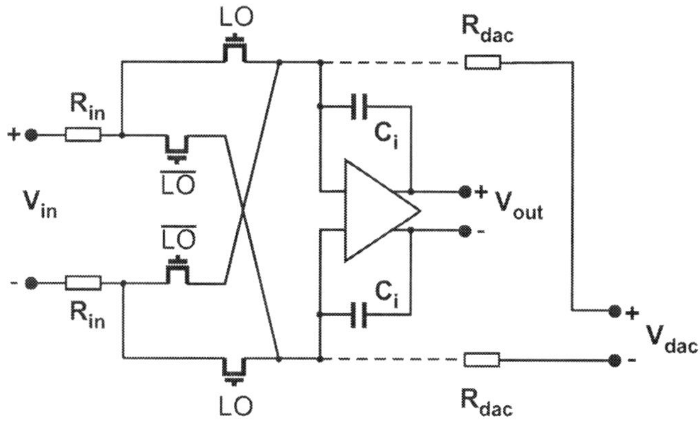

Figure 5-8: *Integrated design of a passive mixer and active RC integrator as a ΣΔ modulator input stage.*

The input resistors and MOS switches implement a linear V/I conversion of the IF input signal. This is possible because the high OTA gain and the feedback configuration create a virtual-ground between the OTA differential input nodes. The virtual-ground nodes then implement a linear summation of the down-converted input and feedback DAC currents. Periodically the switches connect and cross-connect the input resistor to the virtual-ground nodes. In the time domain, the IF input V_{in} is multiplied by a rectangular waveform with frequency f_{LO}. In the frequency domain, the IF input is up and down-converted. The high-frequency mixer products are then filtered-out by the CT loop filter. Because the common-mode levels at the modulator input and at the 1st OTA input are the same, there is no bias current through the mixer switches. The mixer circuit power consumption is therefore negligible.

5.2.1 Mixer Linearity

The linearity of the passive mixer topology in Figure 5-8 is an important design issue. The reception of multi-channel analog signals like AM demands high linearity (IM_3 > 85dB). The analysis presented in [9] assumes that the input resistors R_{in} are ideal and the mixer switches are connected to perfect virtual-ground nodes (the opamp is ideal). The mixer transistors are modelled as ideal switches in series with the non-linear resistance of a MOS transistor biased in the triode region (Figure 5-9).

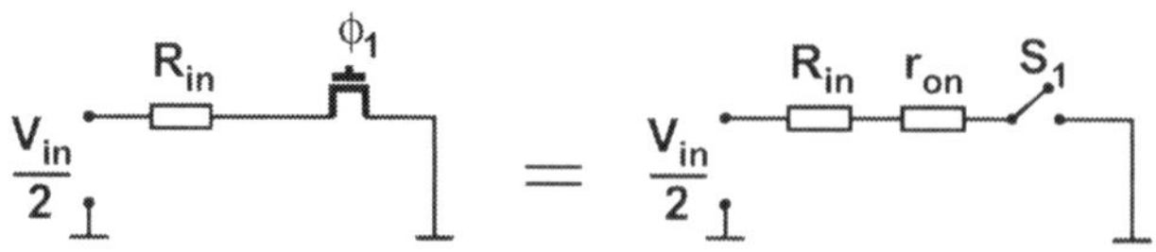

Figure 5-9: *Single-sided mixer model for linearity analysis.*

The non-linear switch resistance $r_{on} = \partial V_{DS}/\partial I_D$ for an NMOS transistor in the triode region is expressed by [10]:

$$r_{on} = \frac{L}{\mu_n C_{ox} W(V_{GS} - V_T - V_{DS})} \tag{5-1}$$

The simple equating of the equivalent circuit in Figure 5-9 gives the basic expressions relating the NMOS switch (V_{DS}, r_{on}) and the input signal voltage and current. Further algebraic manipulation and the use of Taylor series approximation leads to a simplified expression for the 3rd-order harmonic distortion (HD_3) due to the switch non-linearity [9]:

$$HD_3 \approx \frac{3}{32} \cdot \left(\frac{V_{in}}{V_{GT}}\right)^2 \cdot \left(\frac{r_{on}}{R_{in}}\right)^3 \tag{5-2}$$

where $V_{GT}=V_{GS}-V_T$ is the gate overdrive voltage and V_{in} is the maximum input sinusoidal amplitude. The distortion can be minimized if V_{GT} and R_{in} are maximized and large switch transistors (small r_{on}) are used. However, very large switches have larger parasitic capacitances and more significant charge injection. In this design, the input resistors are 1.65kΩ (about 64 times smaller than R_{in} used in the previous generation of car radio IF-to-baseband ADCs [3]). The mixer switches were sized with W/L=60/0.18 (r_{on}=27Ω, V_{GT}=0.33V). As a result, Eq. (5-2) predicts HD_3=114dB due to the non-linearity of the mixer NMOS switches.

5.2.2 Mixer Dynamic Performance

The mixer switches are placed between the input resistors and the input nodes of the opamp. These nodes are also the summing nodes of the ΣΔ modulator, where the differential feedback DAC is connected (Figure 5-8). Every half mixer cycle, the input resistors are connected and cross-connected to the OTA input nodes (Figure 5-10).

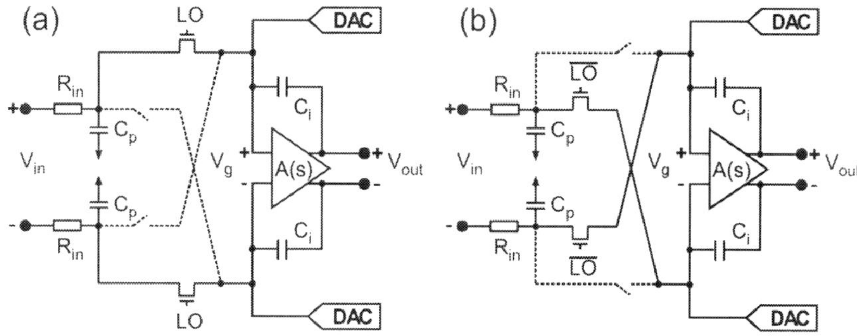

Figure 5-10: *Dynamic behaviour of the MOS mixer with parasitic capacitance C_p, (a) phase* LO, *(b) phase* $\overline{\text{LO}}$.

When LO is high, the parasitic capacitances C_P are charged with the residual virtual-ground voltage V_g. At the beginning of phase $\overline{\text{LO}}$, the parasitic capacitances are reconnected to the OTA inputs and the residual charge flows to the integrating capacitors. C_P presents two main components: the R_{in} poly-to-nwell C_{pN} capacitance, and the mixer switch C_{DB} capacitance. In this design, C_{pN}=130fF and C_{DB}=5fF.

Residual voltage V_g across the OpAmp inputs

A better insight into the effect of the residual charge stored in C_P on the performance of the IF-to-baseband ΣΔ modulator can be obtained if it is expressed as a function of the shaped quantization noise (N_Q). This relation can be calculated with the aid of a simplified model for the 1st integrator shown in Figure 5-11. In this model, the summing node of the ΣΔ modulator is explicitly placed before the input resistors. Because the feedback DAC voltage V_{dac} contains both the input signal V_{in} and N_Q, the 1st integrator input signal can be approximated by N_Q:

$$V_{in} - V_{dac} \cdot R_{in}/R_{dac} \cong N_Q \tag{5-3}$$

The 1st integrator output voltage can be expressed as a function of N_Q:

$$V_{out}(s) = Int_1(s) \cdot N_Q(s) \cong -A_0 / \left(s \cdot \frac{A_0}{\omega_1} + 1\right) \cdot N_Q(s) \tag{5-4}$$

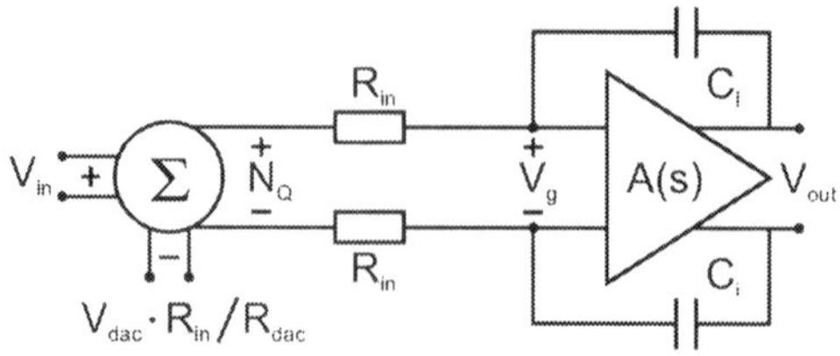

Figure 5-11: *Simplified model for the derivation of V_g as a function of the shaped quantization noise N_Q.*

where A_0 is the opamp dc gain and $\omega_1 = 1/R_{in}.C_i$ is the 1st integrator radian unity gain frequency. The residual voltage across the virtual-ground nodes can be then calculated as function of V_{out} in the frequencies of interest:

$$V_g(s) = \frac{V_{out}(s)}{A(s)} \cong \left(s \cdot \frac{A_0}{\omega_u} + 1\right) / \left(s \cdot \frac{A_0}{\omega_1} + 1\right) \cdot N_Q(s) \qquad (5\text{-}5)$$

where $A(s)$ is the opamp transfer function and $\omega_u >> \omega_1$ is the opamp radian unity gain frequency. This model assumes the feedback signal is integrated continuously. In the case of a SC feedback DAC, the spectral content of $V_g(s)$ can only be predicted by (5-5) at the end of the DAC discharging duty cycle. At this moment, independently of the DAC pulse waveform, all the feedback charge has been integrated and the integrator output has settled to its final value.

Example 5.1: for A_0=90dB, $\omega_1 = 2\pi.1.6$MHz and $\omega_u = 2\pi.650$MHz, the transfer function $V_g(s)/N_Q(s)$ is plotted in Figure 5-12. In the band of interest, V_g is a replica of the shaped quantization noise N_Q attenuated by 52dB.

Sampling of V_g

The residual voltage V_g across the virtual-ground nodes of the 1st integrator are sampled and stored in the parasitic capacitances C_P at the end of the mixer phase LO (Figure 5-10), and discharged to the same nodes with inverted polarity at the beginning of the complementary mixer phase $\overline{\text{LO}}$ (and vice-versa). Therefore, V_g is sampled at twice the mixer switching frequency (f_{LO}). The precise effect of this parasitic sampling depends on the ratio between f_{LO} and f_S. This ratio is normally a power of

2 due to the simplicity of the implementation. In order to be valid to any kind of feedback pulse waveform, the following analysis assumes the mixer LO driver signal only changes its state at the end of the DAC duty cycle. Two examples are discussed in detail:

Case 1: $f_{LO} = f_S/2$

When $f_{LO}=f_S/2$, V_g is sampled at a rate f_S. As a result, the sampled version of V_g contains the same information as the original analog signal without aliasing. The parasitic charge stored in C_P is integrated at the same rate as the error signal (V_{in}-V_{dac}.R_{in}/R_{dac}) coming from the ΣΔ input summing node. Because the spectrum of V_g is an attenuated replica of the error signal spectrum in the frequencies of interest, the output of the 1st integrator is basically unchanged by the sampling and integration of V_g and the noise floor within the signal band is not increased.

Case 2: $f_{LO} = f_S/4$

When $f_{LO}=f_S/4$, V_g is sampled at a rate $f_S/2$ and the spectrum content from $f_S/4$ to $f_S/2$ is aliased. The spectral components located at an arbitrary frequency $f_S/2$ - f_X are aliased to f_X. This means the strong tones near $f_S/2$ will be seen attenuated near dc after the subsampling. Figure 5-13 shows the spectrum of V_g subsampled by a factor 2, that is stored in C_P. A tone with -55dB power can clearly be seen at low frequencies. To better estimate the effect of the subsampling of V_g on the ΣΔ modulator output spectrum, the DT transfer function from C_P to the integrating capacitor C_i has to be calculated.

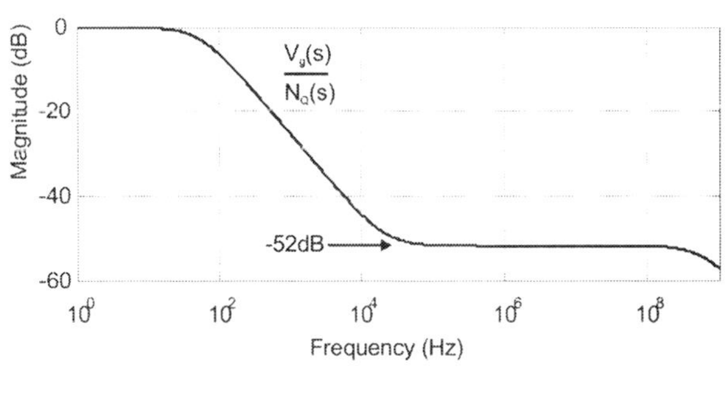

Figure 5-12: *Transfer function* $V_g(s)/N_Q(s)$.

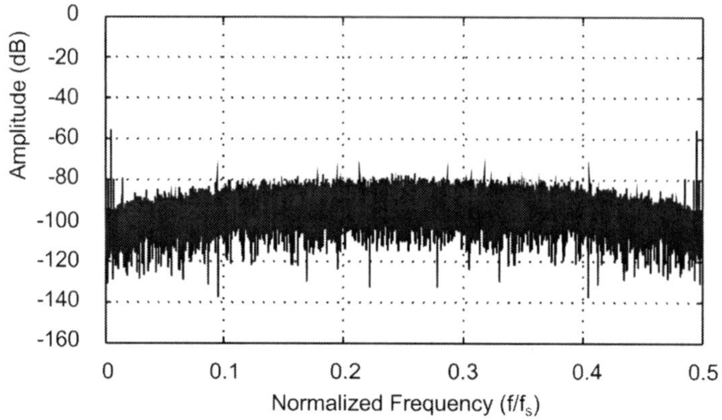

Figure 5-13: *Simulated spectrum (MatLab) of the 2-times subsampled residual voltage V_g stored in C_P.*

Charge transfer

After sampling V_g, the capacitances C_P are cross-connected to the OTA inputs. At this instant, the subsampled quantization noise stored in C_P is transferred to C_i. Figure 5-14 shows the single-ended model of the mixer depicted in Figure 5-8 used to calculate the charge transferred from C_P to C_i. The current $i_g(t)$ flowing through r_{on} is given by:

$$i_g(t) = \frac{v_p(0)}{r_{on}} \cdot e^{\left(\frac{-t}{R_p C_p}\right)} \tag{5-6}$$

where $R_p = r_{on}//(R_{in}+R_s)$ is the equivalent resistance to ground, and R_s is output resistance of the signal source driving the mixer. The charge transfer ends after $T_s/2$ s, when the DAC is connected again to the virtual-ground nodes. Therefore the total charge transferred to C_i is:

$$\Delta Q_i = \frac{R_P}{r_{on}} \cdot C_P \cdot v_p(0) \cdot \left[1 - e^{\left(\frac{-T_S/2}{R_p C_p}\right)}\right] \tag{5-7}$$

Based on Eq. (5-7), the z-domain transfer function of the discrete-time integrator with sampling capacitor C_p can be written:

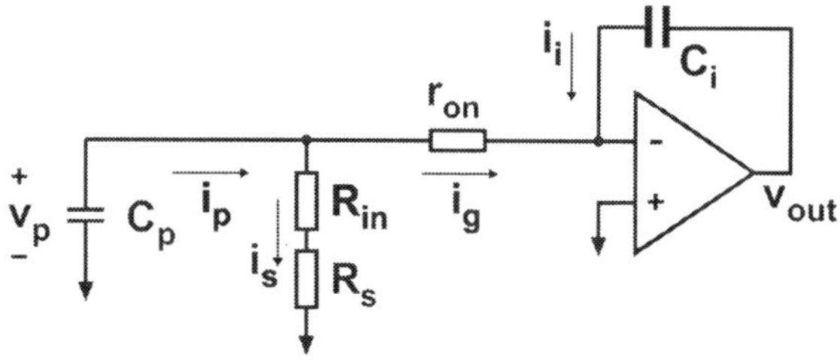

Figure 5-14: *Single-ended model for charge transfer.*

$$\frac{V_{out}(z)}{V_p(z)} = \frac{\frac{C_p}{C_i} \cdot \frac{R_{in} + R_s}{r_{on} + R_{in} + R_s} \cdot \left(1 - e^{\left(\frac{-T_S/2}{R_p C_p}\right)}\right)}{z - 1} \tag{5-8}$$

For frequencies below $f_S/4$, the discrete-time integrator frequency response can be approximated by an equivalent continuous-time integrator. The application of the Euler transformation $s.T_S = z\text{-}1$ in (5-8) leads to (assuming 99% settling and $r_{on} << R_{in} + R_s$):

$$\frac{V_{out}(z)}{V_p(z)} \Rightarrow \frac{V_{out}(s)}{V_p(s)} \cong \frac{C_p / C_i}{s \cdot T_S} \tag{5-9}$$

The input-referred transfer function of the subsampled residual voltage V_p is then approximated by:

$$\frac{V_{in}(s)}{V_p(s)} = \frac{V_{out}(s) / V_p(s)}{Int_1(s)} \cong \frac{C_p / sT_S \cdot C_i}{1 / R_{in} C_i \cdot s} = \frac{R_{in} \cdot C_p}{T_S} \tag{5-10}$$

Example 5.2: for R_{in} =1.7kΩ (*W/L*=12/180), C_p=135fF and f_S=42MHz, Eq. (5-10) equals -40dB and the spectrum around $f_S/2$ is down-converted to baseband multiplied by an attenuation factor F_{db}=-(52+3+40)=-95dB. Figure 5-15 shows the (MatLab) simulated output spectrum (black plot) of the single-bit quadrature 5th-order IF-to-baseband ΣΔ ADC modelled with the traditional mixer topology and $f_{LO}=f_S/4$. High-frequency tones with amplitudes -25dB, -15dB, -15dB and -19dB were down-converted to baseband, respectively, with amplitudes -119dB, -109dB, -109dB and -114dB, well above the expected circuit noise floor (gray plot).

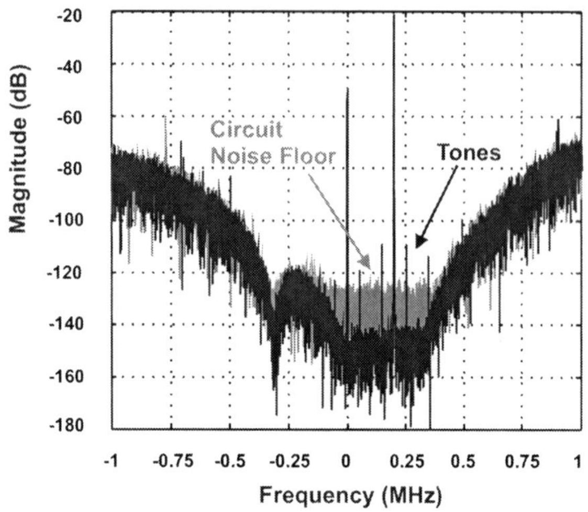

Figure 5-15: *Simulated output spectrum (MatLab) of an IF-to-baseband ΣΔ ADC with traditional mixer topology for $f_{LO}=f_S/4$ and $C_p=135fF$ (RBW=200Hz).*

Experimental evidence of the parasitic effect described in this section was presented in [2]. To reduce the consequences of the quantization noise subsampling, an improved mixer topology was proposed.

5.2.3 Isolated Mixer Topology

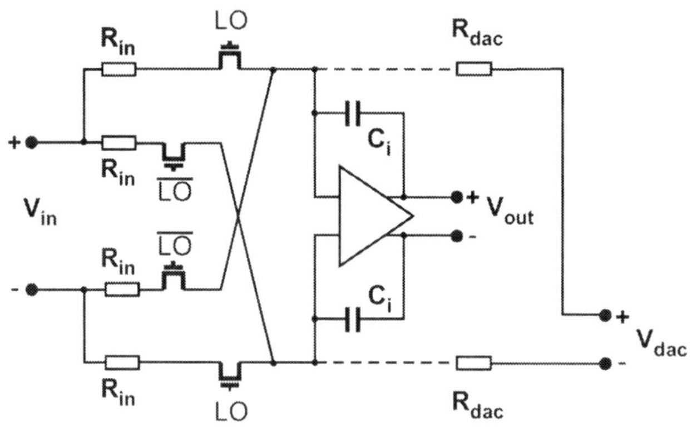

Figure 5-16: *Isolated mixer and active RC integrator.*

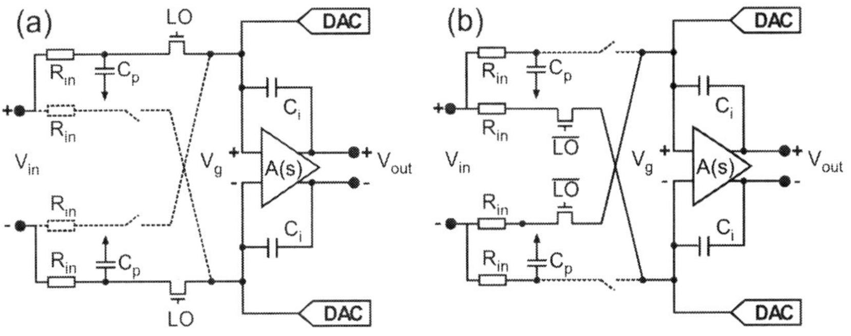

Figure 5-17: *Dynamic behaviour of the isolated MOS mixer with parasitic cap. C_p, (a) phase* LO, *(b) phase* $\overline{\text{LO}}$.

When the ratio f_{LO}=f_S/4 is a system-level requirement, an alternative to reduce the down-conversion of high-frequency tones to the signal band is to change the mixer architecture. Figure 5-16 shows an improved architecture where each switch is connected to its own input resistor [2]. If the signal source preceding the ΣΔ modulator provides a low output impedance, most of the charge accumulated on the parasitic capacitance C_p between R_{in} and the switch transistor does not return to the OTA input nodes. Most of the analysis on the dynamic behaviour of the isolated mixer topology is the same as presented in section 5.2.2. However, the charge transfer analysis is architecture dependent and is developed again in this section.

Charge transfer

As described in Figure 5-17, the virtual-ground voltage V_g is sampled in C_p at the end of the mixer phase LO, and re-connected to the OTA inputs with inverted polarity in the $\overline{\text{LO}}$ phase. Figure 5-18 depicts the single-ended model used to calculate the charge transfer from C_p to C_i when $\overline{\text{LO}}$ is high. The current $i_p(t)$ flowing from C_p is obtained from the equating of the equivalent circuit:

$$i_p(t) = \frac{v_p(0)}{R_x} \cdot e^{\left(\frac{-t}{R_x C_p}\right)} \tag{5-11}$$

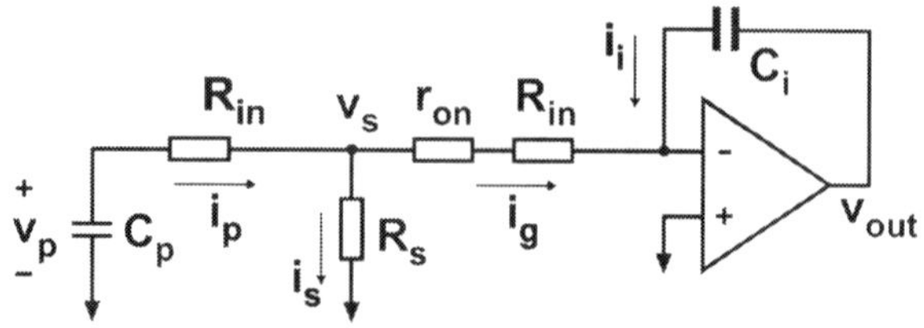

Figure 5-18: *Single-ended model for charge transfer.*

where $v_p(0)$ is the voltage sampled in C_p at the end of the LO phase, R_s is the output resistance of the signal source driving the mixer, and R_x is the equivalent resistance to ground ($R_{in}>>r_{on}$):

$$R_x \cong \frac{R_{in} \cdot (R_{in} + 2R_s)}{R_{in} + R_s} \tag{5-12}$$

The current $i_g(t)$ flowing to the virtual-ground can be obtained from the combination of the expressions for $i_p(t)$, $v_p(t)$ and $v_s(t)$:

$$i_g(t) = \frac{v_p(0)}{R_x} \cdot \frac{R_s}{R_{in} + R_s} \cdot e^{\left(\frac{-t}{R_x C_p}\right)} \tag{5-13}$$

Once again, the total charge transferred from C_p to C_i can be calculated from the integration of $i_g(t)$ during half period of the DAC clock:

$$\Delta Q_i = \frac{R_s}{R_{in} + R_s} \cdot C_P \cdot v_p(0) \cdot \left[1 - e^{\left(\frac{-T_S/2}{R_x C_p}\right)}\right] \tag{5-14}$$

Based on (5-14), the z-domain transfer function of discrete-time integrator with sampling capacitor C_p can be determined for the isolated mixer topology:

$$\frac{V_{out}(z)}{V_p(z)} = \frac{\frac{C_p}{C_i} \cdot \frac{R_s}{R_{in} + R_s} \cdot \left(1 - e^{\left(\frac{-T_S/2}{R_x C_p}\right)}\right)}{z - 1} \tag{5-15}$$

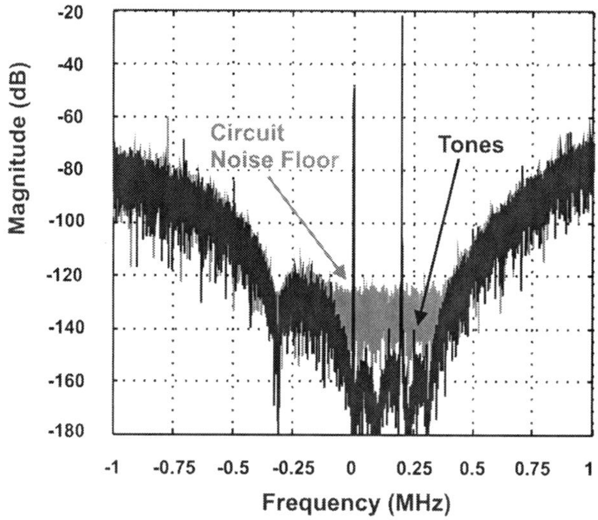

Figure 5-19: *Simulated output spectrum (MatLab) of an IF-to-baseband ΣΔ ADC with isolated mixer topology for* $f_{LO}=f_S/4$ *and* $C_p=135fF$ *(RBW=200Hz).*

Improvement factor

The ratio between Eqs. (5-7) and (5-14) gives the improvement factor *IFR* achieved with the isolated mixer (Figure 5-16) compared with the traditional mixer architecture (Figure 5-8). Assuming 99% settling and $R_{in}>>r_{on}$, the *IFR* can be approximated by:

$$IFR \cong (R_s + R_{in})/R_s \tag{5-16}$$

Example 5.3: for R_{in}=1.7kΩ and R_s=50Ω, the improvement factor is *IFR*=31dB. If the isolated mixer were used for the calculations shown in Example 5.2, the strong tone near $f_s/2$ would be down-converted to the signal band multiplied by an attenuation factor F_{db}=-(95+31)=-126dB. Figure 5-19 shows the (MatLab) simulated output spectrum (black plot) of the single-bit quadrature 5th-order IF-to-baseband ΣΔ ADC modelled with the isolated mixer topology and $f_{LO}=f_S/4$. Low-frequency tones can still be seen above the quantization noise floor (black plot), but they are now well below the expected circuit noise floor (gray plot).

5.2.4 Mixer Driver

As demonstrated in [9], the performance of a ΣΔ modulator integrated with the traditional mixer architecture shown in Figure 5-8 is sensitive to overlapping of the signals LO and $\overline{LO}$ generated by the mixer driver circuit. When both LO and $\overline{LO}$ are logic high, all switches are conducting and the relative integrated current error $\Delta i_i/i_i$ during the overlapping phase can be calculated ($R_{in} >> r_{on}$):

$$\frac{\Delta i_i}{i_i} \cong \frac{1}{1 + G_m r_{on}} \tag{5-17}$$

When all switches are open (LO and $\overline{LO}$ are logic low), the integrated current relative error then becomes:

$$\frac{\Delta i_i}{i_i} \cong \frac{1}{2 G_m R_{in}} \tag{5-18}$$

Behavioural simulations presented in [9] show how the *SNR* of a CT ΣΔ modulator integrated with the standard passive mixer topology (Figure 5-8) is degraded by overlapping mixer driving signals. The relative error $\Delta i_i/i_i$ is maximum when all switches are conducting. The same analysis can be carried out for the isolated mixer topology shown in Figure 5-16. When LO is high and $\overline{LO}$ is low (or vice-versa), as depicted in Figure 5-20a, the expression for the current i_i flowing into the integrating capacitors C_i is given by:

$$i_i = \frac{1}{2R_{dac} + \frac{1}{G_m} \cdot \left(1 + \frac{R_{dac}}{R_{in} + r_{on}}\right)} \cdot V_{dac} \tag{5-19}$$

When all switches are open (Figure 5-20b), the integrated current is expressed by:

$$i_i = \frac{1}{2R_{dac} + \frac{1}{G_m}} \cdot V_{dac} \tag{5-20}$$

The relative integrated current error, when all switches are open, can be calculated combining (5-19) and (5-20) for $R_{in} >> r_{on}$:

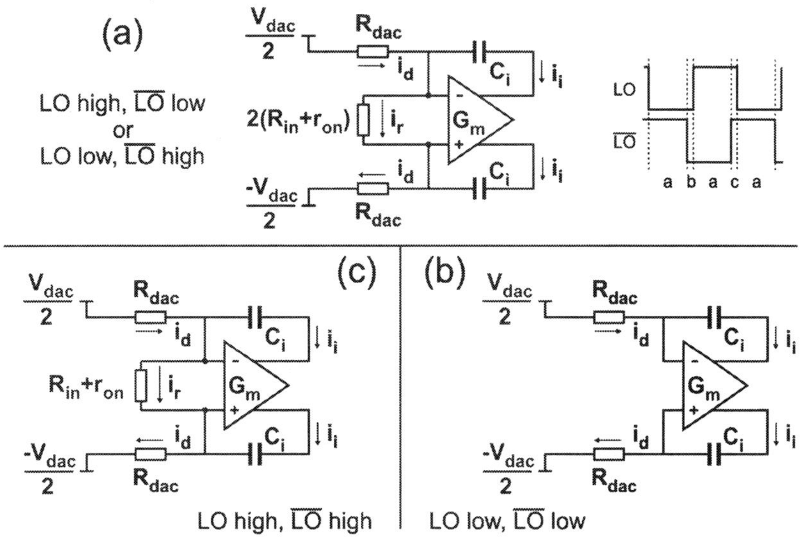

Figure 5-20: *Equivalent circuits for the isolated mixer with overlapping clock phases LO and $\overline{LO}$.*

$$\frac{\Delta i_i}{i_i} = \frac{1}{2G_m R_{in} + 1/\ R_{dac}} \cong \frac{1}{2G_m R_{in}} \tag{5-21}$$

The approximation (5-21) is the same as (5-18). When all switches are closed (Figure 5-20c), the integrated current i_i and the relative error are:

$$i_i = \frac{1}{2R_{dac} + \frac{1}{G_m} \cdot \left(1 + \frac{2R_{dac}}{R_{in} + r_{on}}\right)} \cdot V_{dac} \tag{5-22}$$

$$\frac{\Delta i_i}{i_i} = \frac{1}{R_{in} \cdot (2G_m + 1/\ R_{dac}) + 2} \cong \frac{1}{2(G_m R_{in} + 1)} \tag{5-23}$$

Example 5.4: Assume R_{in} =1.7kΩ and r_{on}=27Ω, and G_m=50mA/V. For the traditional mixer topology in Figure 5-8, the relative errors $\Delta i_i/i_i$ when all switches are off and when all switches are on, are 0.6% and 24% respectively. For the isolated mixer topology (Figure 5-16) the relative error is approximately 0.6% for both situations. This means an improvement of factor of 32dB during the overlapping mixer phase.

5.3 First Integrator

The 1st integrator is the most important loop filter building block, as its noise and non-linearity are directly referred to the input, and normally limits the performance of the entire ΣΔ modulator. Furthermore, the input-referred noise and non-linearity of the following stages are attenuated by the 1st integrator gain. Because the CT IF-to-baseband ΣΔ modulator in this research uses a SC feedback DAC, the 1st OTA was designed according to the requirements of both CT and SC integrators. Figure 5-21 shows the 1st OTA as a CT integrator for the analog input (a), and as a SC integrator for the feedback DAC voltage (b). The mixer, omitted in this picture, is placed between the input resistors (R_{in}) and the OTA inputs.

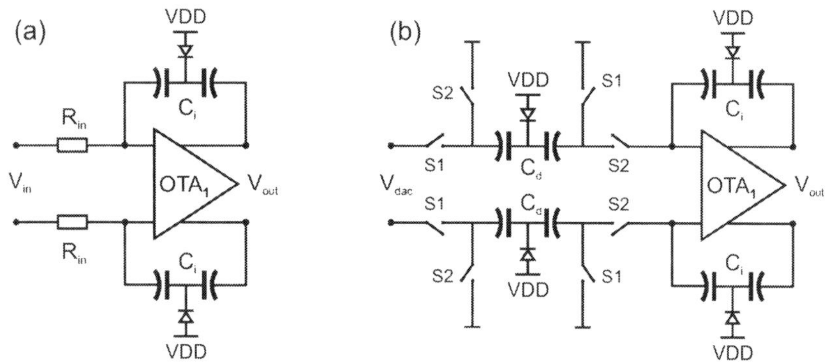

Figure 5-21: *First integrator: CT stage for the modulator input (a), and SC stage for the modulator feedback (b).*

In order to implement an integrator with high linearity, the integrating capacitors C_i are used in a feedback configuration to create a virtual-ground between the OTA inputs. In the 0.18μm CMOS process used to fabricate the prototype ICs, only MOSCAPs and multi-layer metal capacitors are available. Because the latter have a very small capacitance per area (about 20–30 times smaller than the first, as reported in [11]), the MOSCAP option was selected. Figure 5-22 depicts the NMOS-on- nwell capacitor used in this design [7]. Besides the high capacitance per area, MOSCAPs present a capacitance with a large voltage dependence [12]. This device cannot be used as the floating capacitor required by the 1st integrator. A positive voltage large enough to bias the device in the accumulation region is required between the *G* and *N* terminals for linear enough operation. A large parasitic capacitance is also present between the nwell and the substrate.

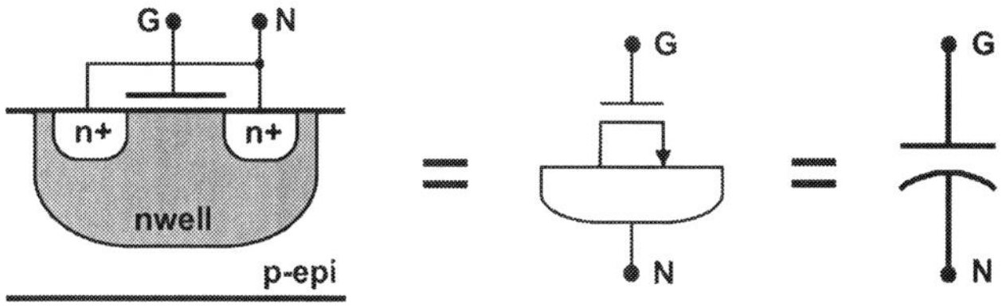

Figure 5-22: *NMOS-on-nwell capacitor [7].*

Series-connected MOSCAPs that can be used as floating capacitors are show in Figure 5-23. The series connection, together with the differential structure of the 1st integrator, provide a 1st-order cancellation of the even-order non-linearities and an attenuation of the odd-order non-linearities if the 2 (4 in a differential structure) capacitors are equal [11]. In order to keep the devices in the accumulation region, the MOS diode pulls the MOSCAP gates to VDD (1.8 V). After the gates are charged to VDD, the diode is turned-off as its V_{GS} becomes zero. The common-mode voltage at the OTA input (set by the ΣΔ modulator feedback) is 0.8 V, while the output common-mode voltage is 0.7 V. Appendix A presents the calculation of the harmonic distortion at the output of a fully differential integrator implemented with MOSCAPs, taking in account the mismatch between the integrating capacitors. The cancellation of the 2nd-order harmonic terms is calculated as a function of the mismatch between the MOSCAPs.

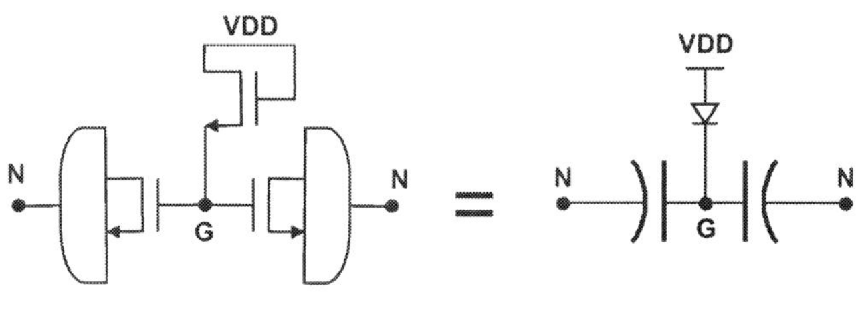

Figure 5-23: *Series-compensated MOSCAPs used as floating capacitor in the 1st integrator and SC DAC.*

5.3.1 Design of the 1st OTA To Be Used as a CT Integrator

The most important design issues regarding the use of the 1st OTA in a CT integrator (Figure 5-21a) are discussed in the following paragraphs. The maximum 0.5 V_{rms} input voltage and the DR requirements in AM mode translate to an input-referred noise floor of -160dB inside the signal band. The resistive input-referred thermal noise of the CT integrator, assuming a RTZ DAC with R_{dac} resistors is expressed by [9]:

$$S_{INref}(f) = 8k\theta R_{in} \cdot \left(1 + \frac{R_{in}}{R_{dac}}\right) \tag{5-24}$$

where k is the Boltzman's constant and θ is the absolute temperature in Kelvin. As demonstrated in Appendix B, Eq. (5-24) is a valid approximation when a SC DAC replaces the RTZ DAC. In this case, $R_{dac}=T_S/C_d$ is the equivalent resistor of a SC network if the OSR is high. The use of (5-24) as an approximation also requires the noise contribution of the 1st OTA to be below the contribution of R_{in} and the DAC's reference voltage to be designed with the output noise below the R_{dac} thermal noise level. The proper scaling of the ratio between the maximum input signal current and the feedback DAC current imposes another relation between R_{in} and R_{dac} [9]:

$$\frac{R_{in}}{R_{dac}} = \frac{1}{1-\delta} \cdot \frac{V_{in}}{V_{dac}} \tag{5-25}$$

where V_{in} is the rms maximum input sinusoidal, V_{dac} is the DAC differential voltage and δ is the DAC return-to-zero duty cycle. In this design, δ=0.5 as the settling time available for the SC DAC charge transfer is $T_S/2$. The combination of (5-24) and (5-25) leads to R_{in}=1.7kΩ and R_{dac}=5.4kΩ. To implement the integrator unity gain frequency f_I=1.7MHz, large integrating capacitors are required (C_i=55pF). Assuming an opamp model with finite dc gain A_0 and unity gain frequency $\omega_u=2\pi . f_u$, the CT transfer function of the 1st integrator is:

$$\frac{V_{out}(s)}{V_{in}(s)} = \frac{1}{s^2 \cdot \frac{R_{in}C_i}{\omega_u} + s \cdot \left(\frac{1}{\omega_u} + \frac{R_{in}C_i}{A_0} + R_{in}C_i\right) + \frac{1}{A_0}} \tag{5-26}$$

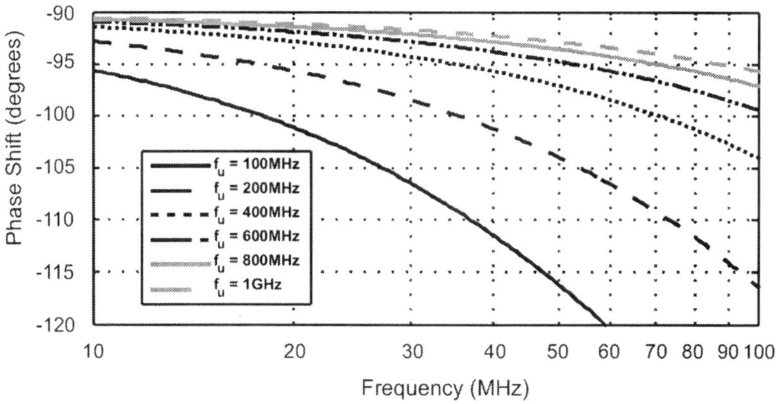

Figure 5-24: *CT 1st integrator phase-shift for several values of the opamp unity gain frequency (f_u).*

The phase-shift of the 1st integrator, for several values of f_u, is plotted in Figure 5-24. Due to the feedforward compensated loop filter architecture (Figure 5-3), the loop filter phase-shift at high frequencies is dominated by the 1st integrator. In order to implement a stable CT loop filter, the 1st integrator phase-shift should be as close as possible to -90deg. In order to satisfy this criteria, the 1st OTA should be designed with $f_u \geq 400$MHz.

The 1st OTA can also be modelled as a finite transconductance G_m. When this model is used to calculate the CT transfer function of the 1st integrator, a right-half-plane (RHP) zero appears in the numerator of the transfer function. As a result, an undesirable extra phase-shift is added to the loop filter phase-shift. In order to cancel the RHP zero, a small resistor R_s can be placed in series with C_i. The transfer function of the compensated OTA then becomes:

$$\frac{V_{out}(s)}{V_{in}(s)} = \frac{sC_i \cdot \left(\frac{1}{G_m} - R_s\right) - \frac{1}{G_m}}{sC_i \cdot \left(\frac{1}{G_m} + R_{in}\right)} \tag{5-27}$$

if $R_s = 1/G_m$, the RHP is cancelled and the extra phase-shift is eliminated.

The transconductance G_m of the 1st OTA should satisfy a linearity and a noise criteria. As demonstrated in [9] assuming a RTZ DAC, the 3rd-order harmonic distortion for an RC feedback integrator inside a ΣΔ modulator is expressed by:

$$HD_3 \cong \frac{V_{in}^2}{64 \cdot G_m \cdot R_{in}^3 \cdot I_b^2} \cdot \left(1 + \frac{R_{in}}{R_{dac}}\right) \quad (5\text{-}28)$$

where I_b is the 1st OTA bias current. For the parameters in this design, HD_3>130dB is already achieved with I_b=20mA and G_m=50mA/V. The input-referred noise of the 1st OTA should be below the resistors thermal noise level. For a single-stage telescopic OTA as depicted in Figure 5-32, the input-referred noise PSD at low frequencies is approximated by [10]:

$$S_{OTA1} \cong \frac{R_{dac} + R_{in}}{R_{dac}} \cdot \left[\frac{16k\theta}{3 \cdot g_{m1}} \cdot \left(1 + \frac{g_{m7}}{g_{m1}}\right) + \frac{2K_P/f}{(WL)_1} + \frac{g_{m7}^2}{g_{m1}^2} \cdot \frac{2K_N/f}{(WL)_7}\right] \quad (5\text{-}29)$$

where g_{m1} and g_{m7} are the transconductances of the OTA input and current source transistors, WL is the product of the transistor dimensions and, K_P and K_N are a process dependent constants that are different for PMOS and NMOS transistors [10]. The ratio $(R_{dac}+R_{in})/R_{dac}$ is the transfer function from the OTA inputs to the integrator inputs at low frequencies. The last two terms in (5-29) account for the $1/f$ noise contributions. In order to obtain the input-referred OTA thermal noise 20dB below the resistors thermal noise, assuming the worst case situation $g_{m7}=g_{m1}$, the OTA transconductance should be at least $G_m = g_{m1}$=60mA/V.

In a ΣΔ modulator, the finite dc gain of the loop filter OTAs limits the *NTF* attenuation at low frequencies. Figure 5-25 shows the frequency response of the complex *NTF* for the CT loop filter used in this design. The several parametrized plots are a function of the 1st OTA dc gain A_0 and the dc gain of the high-order integrators (A_h). The ideal *NTF* (gray plot) is also shown for comparative purposes. A_h is assumed to be 40dB, as it is common for the high-order integrators to be implemented with low dc gain OTAs. Because of the 5th-order loop filter, it is possible to achieve high quantization noise attenuation inside the signal band for values of A_0 greater than 60dB. Even if A_0=40dB, the quantization noise level would stay below the thermal noise floor.

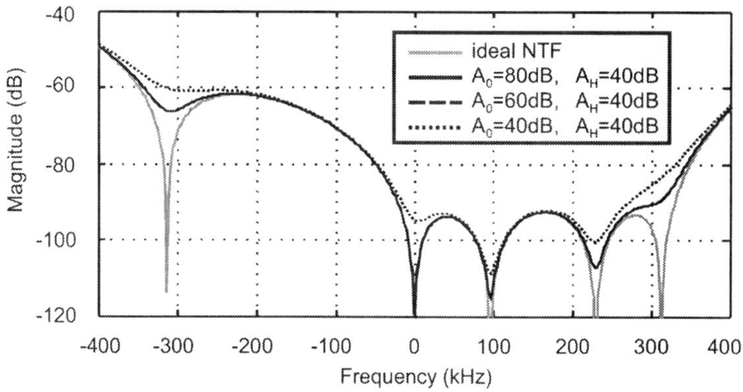

Figure 5-25: *Complex 5th-order NTF frequency response for different values of the 1st OTA dc gain A_0.*

5.3.2 Design of the 1st OTA To Be Used as a SC Integrator

The most important design issues regarding the use of the 1st OTA as part of a SC integrator (Figure 5-21b) are discussed in the following paragraphs. The z-domain transfer function of the SC integrator can be obtained from the charge transfer equations of a SC network. Figure 5-26 shows the single-ended model of the SC integrator used for charge transfer calculations.

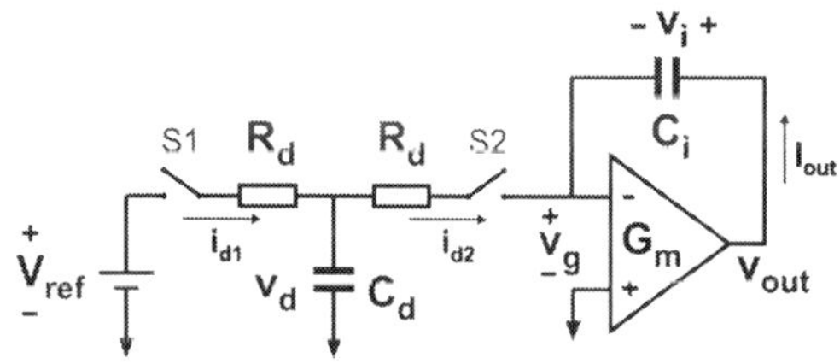

Figure 5-26: *Single-ended model of the SC integrator.*

The solution of the differential equations describing the above circuit during phases S1 and S2 leads to the expressions for the charging and discharging currents $i_{d1}(t)$ and $i_{d2}(t)$.The first is trivial, while the second is given by:

$$i_{d2}(t) = \frac{V_{REF}}{R_d} \cdot \left(1 - e^{\frac{-T_S/2}{R_d C_d}}\right) \cdot e^{\frac{-t}{\left(R_d + \frac{1}{G_m}\right)C_d}} \tag{5-30}$$

Equation (5–30) reveals that the time constant $\tau_2=(R_d+1/G_m)C_d$ during phase S2 is also a function of the OTA transconductance. In order to have the charge transfer determined only by R_d and C_d, the 1st OTA should be designed for $G_m>>1/R_d$. If this condition is satisfied, the SC integrator transfer function is approximated by:

$$\frac{V_{out}(z)}{V_{dac}(z)} \cong \frac{\frac{C_d}{C_i} \cdot \left(1 - e^{\frac{-T_S/2}{R_d C_d}}\right)^2}{z - 1} \tag{5-31}$$

Example 5.5: In order to minimize the effect of the transconductance on the settling behaviour of the current $i_{d2}(t)$, $1/G_m$ is chosen to be 5% of R_d. For R_d =265Ω, the OTA transconductance should be G_m=75mA/V.

Due to the SC DAC discharge, the voltage across the virtual-ground nodes can experience large transients every time the DAC capacitors are connected to the OTA inputs. As long as the settling is linear, as described in Eq. (5-30), the only effect on the SC integrator transfer function is a gain error caused by the incomplete charge transfer. In a SC ΣΔ modulator, only the final values of the internal loop filter states are relevant. However, in a CT ΣΔ modulator, the instantaneous values of the internal states are important as the integrators process the analog signals continuously. Therefore, the dynamic behaviour of the voltage across the OTA inputs may affect the linearity of the 1st integrator and should be properly quantified. Based on the model shown in Figure 5-26, a closed form expression for $v_g(t)$ in phase S2 can be calculated:

$$v_g(t) = \frac{v_d(0)}{G_m \cdot R_d} \cdot e^{\frac{-t}{\left(R_d + \frac{1}{G_m}\right)C_d}} \tag{5-32}$$

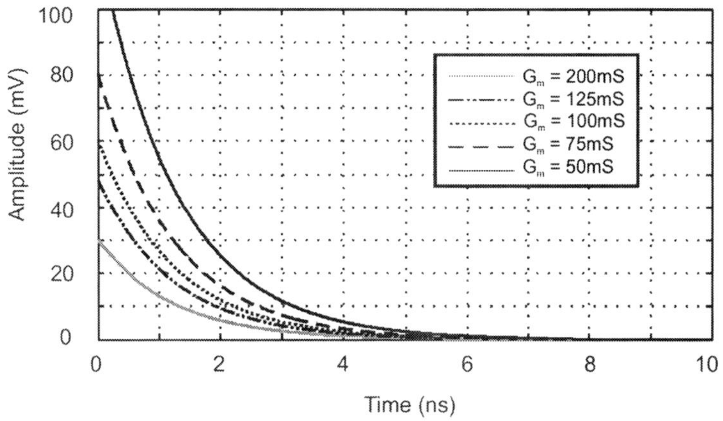

Figure 5-27: *Virtual-ground $v_g(t)$ dynamic behaviour for several values of the OTA transconductance G_m.*

Figure 5-27 shows the dynamic behaviour of the virtual-ground voltage after a SC DAC discharge for several values of the OTA transconductance. For G_m = 50 mS, the peak value of $v_g(t)$ is greater than 120 mV (240 mV differential)!

Example 5.6: For R_d=265 Ω and $v_d(0)$=1.6 V, G_m should be greater than 130 mS in order to have the peak value of $v_g(t)$ below 50 mV. This example demonstrates that peak voltage at the OTA input and not the input-referred thermal noise, nor the linearity as expressed in Eq. (5-28), is the dominant effect on the determination of the required transconductance for the 1st OTA.

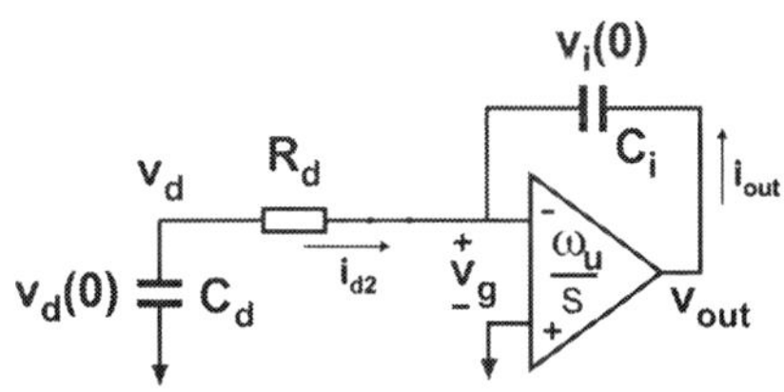

Figure 5-28: *Single-ended model of the SC integrator in phase S2, with opamp transfer function $A(s)=\omega_u/s$.*

The finite unity gain frequency $\omega_u=2\pi.f_u$ of the 1st OTA also impacts the settling behaviour of the current $i_{d2}(t)$. In order to determine the charge transfer in this case, the OTA is modelled as an opamp with voltage gain $A(s)=\omega_u/s$. Figure 5-28 shows the equivalent single-ended model of the SC integrator during phase S2. In this case, the solution of the differential equation for $i_{d2}(t)$ can be approximated by:

$$i_{d2}(t) \cong \frac{v_d(0)}{R_d} \cdot \left(\frac{C_d}{C_i} \cdot e^{-\omega_u t} + e^{\frac{-t}{R_d C_d}} \right) \tag{5-33}$$

When the dynamic behaviour of the OTA is accounted, $i_{d2}(t)$ is characterised by two independent settling components. To obtain the settling of $i_{d2}(t)$ dominated by the R_dC_d time constant, the 1st OTA should be designed for $1/\omega_u << R_dC_d$. Based on the model shown in Figure 5-28, another closed form expression for $v_g(t)$ can be obtained after some approximations:

$$v_g(t) \cong \frac{v_d(0)}{(R_d C_d \omega_u - 1)} \cdot \frac{C_d}{C_i} \cdot \left(e^{\frac{-t}{R_d C_d}} - e^{-\omega_u t} \right) \tag{5-34}$$

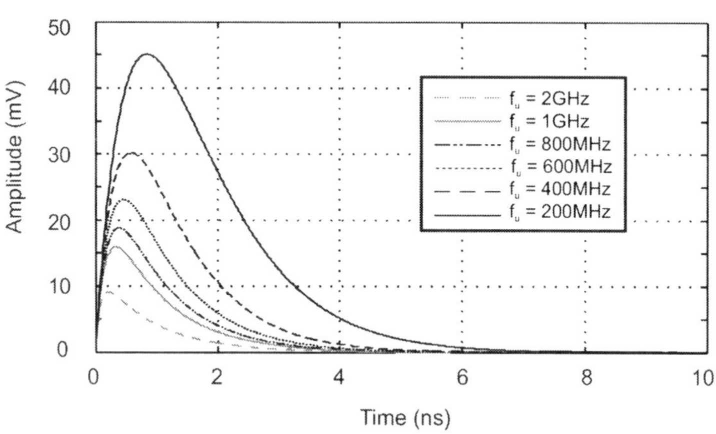

Figure 5-29: *Virtual-ground voltage $v_g(t)$ dynamic behaviour for several values of the OTA unity gain frequency f_u and infinite dc gain.*

Figure 5-29 shows the dynamic behaviour of the virtual-ground voltage for several values of the 1st OTA unity gain frequency. The peak value of $v_g(t)$ happens when $dv_g(t)/dt=0$. The instant t_{max}, when $v_g(t)$ is peaking, can be then calculated:

$$t_{max} = \frac{-R_d C_d \cdot \ln(R_d C_d \omega_u)}{1 - R_d C_d \omega_u} \tag{5-35}$$

The substitution of (5-35) into (5-34) leads to an expression for $v_g(t_{max})$:

$$v_g(t_{max}) = \frac{C_d}{C_i} \cdot v_d(0) \cdot (R_d C_d \omega_u)^{\frac{R_d C_d \omega_u}{1 - R_d C_d \omega_u}} \cong \frac{v_d(0)}{R_d C_i \omega_u} \tag{5-36}$$

Equation (5–36) reveals that the peaking across the 1st OTA input nodes is inversely proportional to the unity gain frequency ω_u.

Example 5.7: According to (5-36), for R_d=265Ω, C_i=55pF and $v_d(0)$=1.6V, f_u should be greater than 400MHz in order to have the peak value of $v_g(t)$ below 30mV.

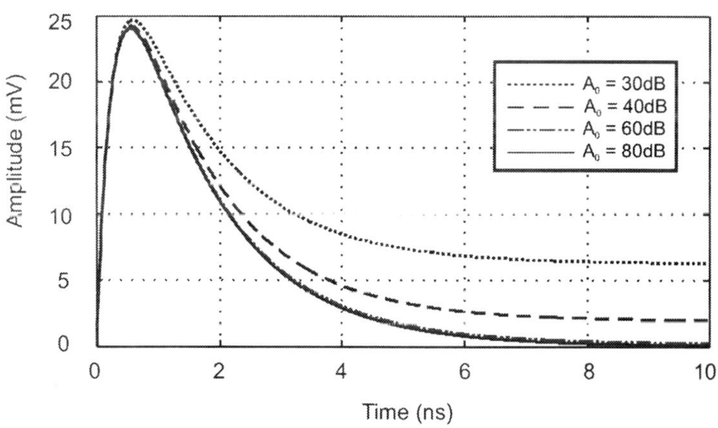

Figure 5-30: *Virtual-ground voltage $v_g(t)$ dynamic behaviour for several values of the OTA dc gain A_0 and f_u=600MHz.*

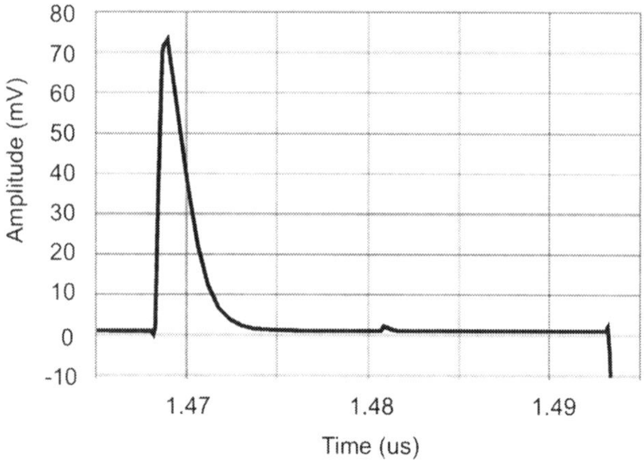

Figure 5-31: *Differential virtual-ground voltage, PSTAR circuit simulation for $G_m=120Ms$ and $f_u=610MHz$.*

To also take in account the effect of the OTA dc gain A_0 in the determination of $v_g(t)$, the transfer function of the opamp in Figure 5-28 should become $A(s)=1/(1/A_0+\omega_u/s)$. In this case, the s-domain expression for the virtual-ground voltage $V_g(s)$ is given by:

$$V_g(s)=\frac{(s+\omega_u/A_0)\cdot[C_iv_i(0)-C_dv_d(0)+s\cdot C_dC_iR_dv_i(0)\omega_u]}{s^3\cdot C_dC_iR_d+s^2\left[C_d+C_i+C_dC_iR_d\omega_u\left(1+\frac{1}{A_0}\right)\right]+s\cdot\omega_u\left(\frac{C_d+C_i}{A_0}+C_i\right)} \quad (5\text{-}37)$$

The application of the inverse Laplace transformation over (5-37) does not result into a simple analytical expression for $v_g(t)$. Figure 5-30 shows numerical solutions of $v_g(t)$ for f_u=600MHz and some values of A_0. The finite dc gain A_0 only significantly affects the final value of $v_g(t)$. For A_0>60dB, the final value of $v_g(t)$ is already much smaller than 1mV.

The actual dynamic behaviour of the differential virtual-ground voltage after the SC DAC discharge is depicted in Figure 5-31. It was obtained from a complete PSTAR ΣΔ circuit simulation with the OTA shown in Figure 5-33 and the parameters used in this design. The single-ended peak virtual-ground voltage is $v_g(t_{max})$=38mV. Figure 5-31 reveals that the actual $v_g(t)$ transient is in between the dynamic behaviours predicted by Eqs. (5-32) and (5-34). For this design, the finite OTA G_m is the most important parameter to estimate the peaking of $v_g(t)$. The discrepancy between (5-36) and the PSTAR circuit simulation can be explained in part by the fact that the nominal switch resistance r_{on} increases when $v_g(t)$ is peaking, as the V_{GS} of the DAC switch transistor is reduced. The DAC resistor R_d is actually composed of a linear poly-resistor R_d' =85 Ω and 2 times r_{on} (section 5.6).

5.3.3 1st OTA Transistor Level Design

Figure 5-32 shows the single-stage telescopic OTA [13] designed for the first prototype IC. All transistors are stacked in two differential paths, which limits the differential output swing to 0.4 V in order achieve the required linear performance of the output transistors.

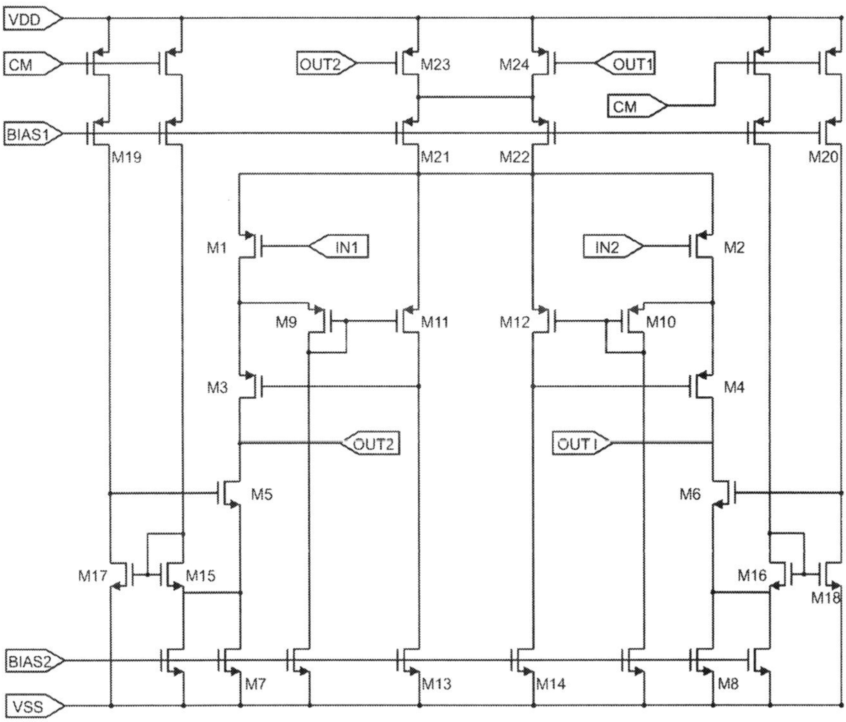

Figure 5-32: *First OTA, first prototype.*

The input differential pair M1-M2 provides a high transconductance g_{m1} (110mA/V). PMOS input transistors are used in order to obtain a lower *1/f* noise. The transconductance g_{m7} of the current sources M7-M8 was designed to be 20% smaller than g_{m1}. In order to minimize the *1/f* noise contribution of the NMOS current sources M7-M8, these transistors were designed with L=1µm, 5 times the minimum transistor length. The obtained input-referred *1/f* noise corner frequency was about 100kHz.

The cascode transistors M3-M4 and M5-M6 keep the output impedance high. However, just 60dB of dc gain could be obtained from a cascoded single-stage telescopic amplifier. To achieve a higher dc gain in a 0.18µm CMOS process, gain boosted cascodes were used [14]. The level-shifter transistors M9-M10 sense the drain voltage of M1-M2. The gain boosting amplifiers M11-M13 and M12-M14 control the gate voltage of the cascodes M3-M4 in order to keep the V_{DS} of M1-M2 fixed. The gate voltages of M5-M6 are controlled in the same way by the amplifiers M17-M19 and M18-M20. Special attention was paid to the design of the gain boosted amplifiers in order to avoid slow settling components. In order to implement the high transconductance, a bias current I_b=40mA was required for this OTA. The 1st OTA accounts for 70% of the total power consumption of the loop filter. Its bias current is high enough to prevent slewing due to the SC DAC discharge even for the worst-case DAC peak-current of 8mA. The dc gain is 102dB and the unity gain frequency f_u is 380MHz. The triode biased transistors M23-M24 are employed on the common-mode control circuitry [13]. The stability of the common-mode control loop, as well the stability of the gain boosting loops, was checked via ac and transient simulations.

Figure 5-33 shows the 1st OTA circuit designed for the second prototype. Again it is a single-stage telescopic OTA with PMOS input transistors M1-M2 and gain boosted cascodes M3-M4 and M5-M6 to achieve high dc gain. The common-mode control is implemented with the error amplifier M30-M36 that compares a reference level with the actual CM level generated by the average of the two differential outputs. The capacitive load connected to the differential output is smaller compared to the first prototype. For this second design, the dc gain was reduced to 95dB and the unity gain frequency increased to 610MHz. The bias current remained I_b=40mA and the achieved G_m was 120mA/V.

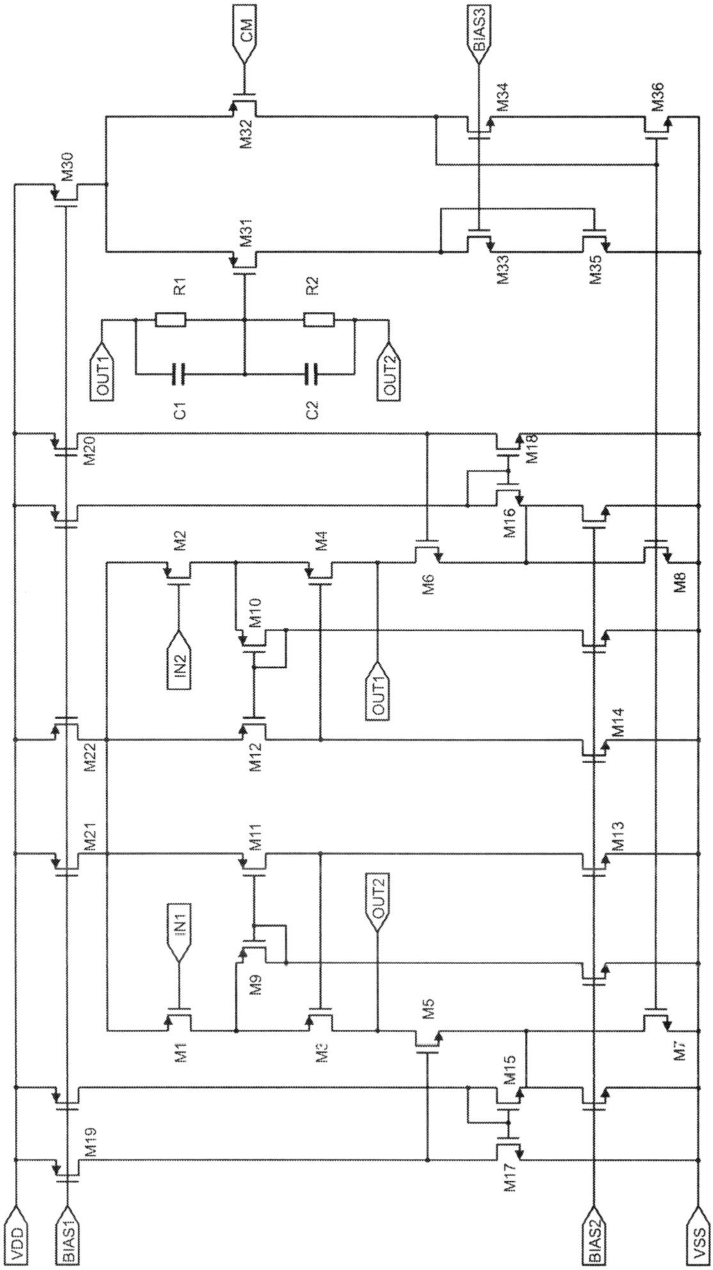

Figure 5-33: *First OTA, second prototype.*

5.4 High-Order Integrators and Resonators

The loop filter high-order integrators were implemented with high input impedance G_m-C stages (Figure 5-34). The input-referred noise of these stages is attenuated by the gain of the preceding ones. Figure 5-35 shows the transfer functions from each OTA input to the loop filter input. The input-referred thermal noise contribution from each loop filter building block is designed to be below the input resistors thermal noise floor. The input-referred non-linearities of the high-order integrators are also attenuated by the preceding stages. A minimum HD_3=-70dB, determined from behavioural simulations, is required from the 2nd OTA.

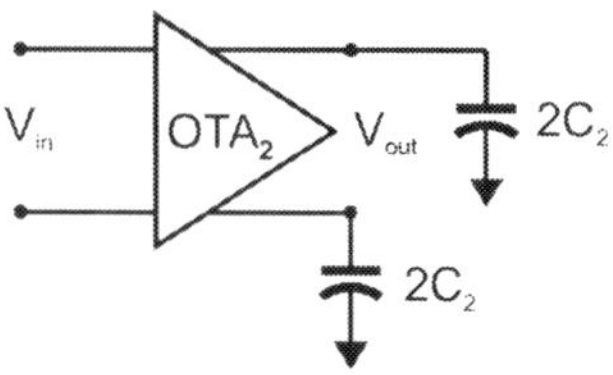

Figure 5-34: *G_m-C integrator with MOSCAPs.*

In order to achieve the required linearity without feedback and to better define the OTA transconductance G_m, source degeneration is applied to the input transistors of the high-order OTAs. Figure 5-36 shows the OTA used in the G_m-C integrators and resonators. The input pair M1-M2 and the degeneration resistors R_{dg} convert the input voltage into a current, increasing the linear input differential range to 0.4 V. Because the input and the output common-mode levels are the same (0.7 V), this OTA was implemented as a folded cascode amplifier [13]. The current sources M3-M4 sink the current from the input differential branch and from the output branch. The cascodes M5-M6 and M7-M8 provide a high output impedance. A dc gain of about 45 dB is sufficient because the gain of all integrator stages add to the total loop gain. The triode biased transistors M11-M12 control the output common-mode level, set by a replica biasing circuit and the transistors M13-M14. The transconductance of the degenerated OTA is given by:

$$G_m = \frac{1}{1/\ g_{m1} + R'_{dg} + R'_{dg}/\ (g_{m1} \cdot\ r_{o1})} \cong \frac{1}{R'_{dg}} \qquad (5\text{-}38)$$

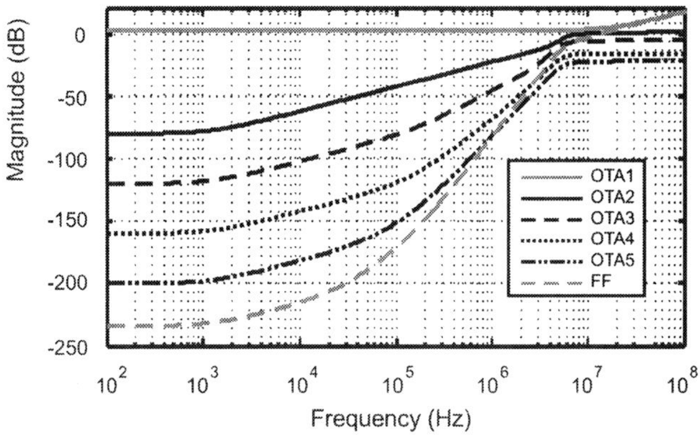

Figure 5-35: *Input-referred transfer functions for each loop filter building block.*

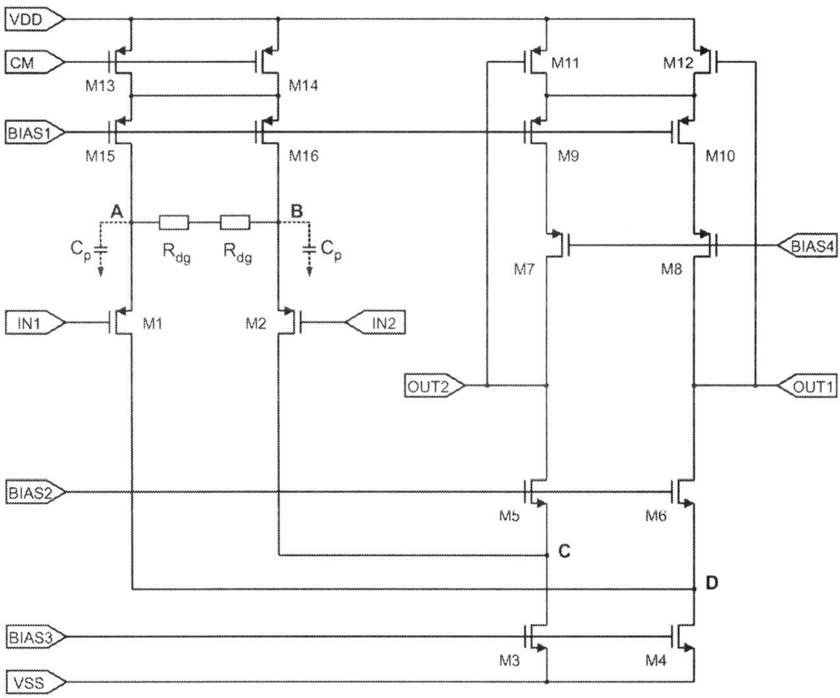

Figure 5-36: *Degenerated folded cascoded OTA.*

where g_{m1} and r_{o1} are, respectively, the transconductance and the output resistance of M1, and $R'_{dg}=R_{dg} // r_{o15}$ is the parallel combination of the degeneration resistor and the output resistance of M15. Equation (5–38) reveals that the higher g_{m1} is, the closer to $1/R'_{dg}$ the OTA transconductance becomes and more linear the OTA behaves. However, the potential increase in linearity is limited by the non-linear output resistance of M15. For high values of the bias current (I_b>2mA) and g_{m1}, the linearity of R'_{dg} is limited by r_{o15}. The input-referred thermal noise from the OTA show in Figure 5-36 can be calculated as a function of its internal components:

$$S_{OTA}=2\left[S_{M1}+S_R\cdot\left(\frac{R'_{dg}}{R_{dg}}\right)^2+(R'_{dg})^2\cdot\left(g_{m3}^2\cdot S_{M3}+g_{m9}^2\cdot S_{M9}+g_{m15}^2\cdot S_{M15}+\frac{S_{M7}}{r_{o9}^2}+\frac{S_{M5}}{r_{o3}^2}\right)\right] \quad (5\text{-}39)$$

where $S_{Mi}=8k\theta/3g_{mi}$ and $S_R=4k\theta R_{dg}$ are the thermal noise contributions from transistor Mi and the degeneration transistor R_{dg}. When the transistor transconductances are defined as $g_{mi}=g_{m1}\cdot\alpha_i$, (5-39) can be rewritten:

$$S_{OTA}\cong 8k\theta(R'_{dg})^2\cdot\left[\frac{1}{R_{dg}}+\frac{2g_{m1}}{3}\cdot\left(\frac{1}{(g_{m1}R'_{dg})^2}+\alpha_3+\alpha_9+\alpha_{15}\right)\right] \quad (5\text{-}40)$$

Equation (5–40) reveals that, for each value of R_{dg}, there is an optimum value of $g_{m1}\cong 1/(R_{dg}.\sqrt{\alpha_3})$ that minimizes the OTA input-referred noise. Figure 5-37 shows S_{OTA} as a function of g_{m1}, for several values of R_{dg}.

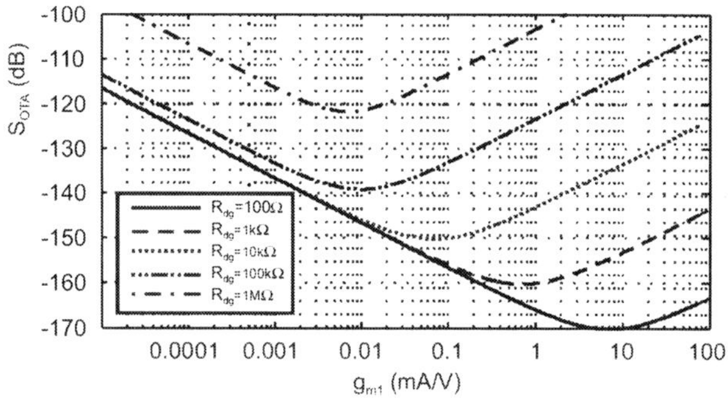

Figure 5-37: *OTA input-referred thermal noise as a function of* g_{m1} *for* $\alpha_i=1$ *and* $r_{o15}=100k\Omega$

The plots shown in Figure 5-37 were obtained for α_i=1 and r_{o15}=100kΩ. Normally, α_9<1 and α_{15}<1 while α_3>1 as M3-M4 carry twice the current of the other devices. Because M3 and M4 are NMOS transistors, they were designed with L=4μm (about 20 times the minimum L) in order to obtain a low enough $1/f$ noise corner frequency for OTA_2. The value of R_{dg} for every OTA was primarily chosen based on the coefficients of the ΣΔ modulator and on the noise transfer functions shown in Figure 5-35. After the resistors R_{dg} were defined, the OTA bias current I_b and g_{m1} were designed to minimize the OTA thermal noise as depicted in Figure 5-37 and to satisfy the minimum required HD_3 level. Table 5-2 shows the total current consumption of the high-order OTAs.

Table 5-2. *High-order OTAs current consumption.*

	I_b (mA)	$2R_{dg}$ (kΩ)
OTA2	4.0	1.5
OTA3	1.6	3.0
OTA4	0.32	24
OTA5	0.32	24
OTAR	0.32	108

The parasitic capacitance C_p, from the nodes A and B (Figure 5-36) to the ground, impact the frequency response of the G_m-C integrator. The non-ideal integrator transfer function is expressed by:

$$Int(s) = \frac{sC_pR'_{dg} + 1}{s\left(\frac{sC_pC_iR'_{dg}}{g_{m1}} + \left(\frac{1}{g_{m1}} + R'_{dg}\right)C_i\right)} \tag{5-41}$$

Equation (5-41) reveals that an extra left-half-plane (LHP) zero is added, together with a high-frequency pole, to the integrator transfer function. The value of the parasitic capacitance C_p is determined mainly by the R_{dg} poly-to-nwell capacitance. It should be minimized in order to obtain the zero frequency above $f_S/2$. In special, the extra integrator zero affects severely the frequency response of the G_m-C resonators. Figure 5-38 shows the (real) continuous-time resonator used in this design.

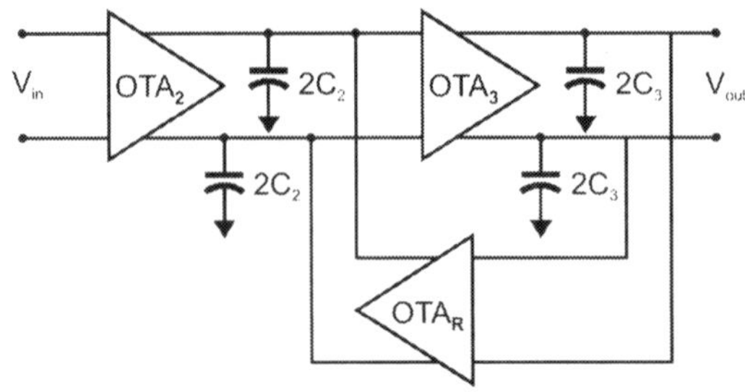

Figure 5-38: *G_m-C (real) resonator with MOSCAPs.*

When C_p is taken in account, the resonator transfer function becomes:

$$\frac{V_{out}(s)}{V_{in}(s)} \cong \frac{s \cdot dC_3R_3 \cdot (1+sC_{p2}R_2)}{s^2 \cdot (dR_2R_3C_2C_3+d^2R_2R_3C_{p2}C_{p3})+s \cdot (R_3C_{p3}+d^2R_2C_{p2})+1} \qquad (5\text{-}42)$$

where C_j, C_{pj} and R_j are, respectively, the integrating capacitor, the parasitic capacitor and the degeneration resistance of the jth OTA. For the feedback OTA$_R$, $R_r=d.R_2$ and $C_{pr}=d.C_{p2}$. The resonance frequency is then $\omega_R=(1/dR_2R_3C_2C_3)^{1/2}$.

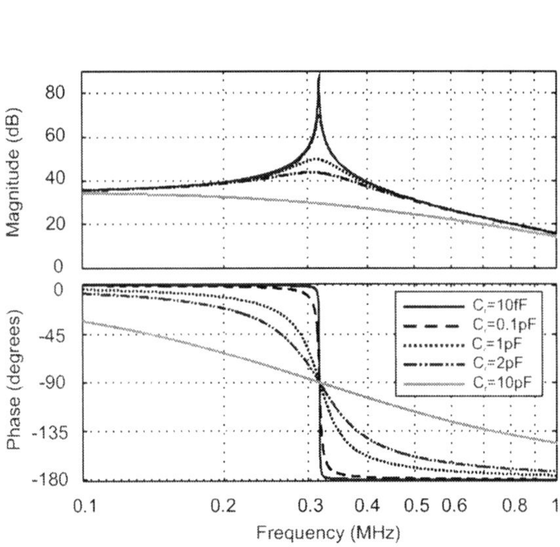

Figure 5-39: *G_m-C resonator Bode plots for several values of the parasitic capacitance $C_{pr}=dC_{p2}$.*

Figure 5-39 shows the Bode plots for the G_m-C resonator described by Eq. (5-42). The effect of the OTAs parasitic capacitances C_{p2}, C_{p3} and $C_{pr}=d.C_{p2}$ is to smooth the resonator frequency response and reduce the magnitude of the resonance. When the parasitic capacitances are too large, the resonant peak vanishes and the expected *NTF* notch is suppressed. As a result, the quantization noise power increases inside the signal band. To prevent this effect, the degeneration resistors should be designed with minimum area in order to minimize the parasitic capacitances to ground.

5.5 Feedforward Coefficients and Quantizer

The feedforward coefficients are implemented as MOSFET-only degenerated transconductors [15] with current output (Figure 5-40). They connect the output of each integrator to the current input quantizer.

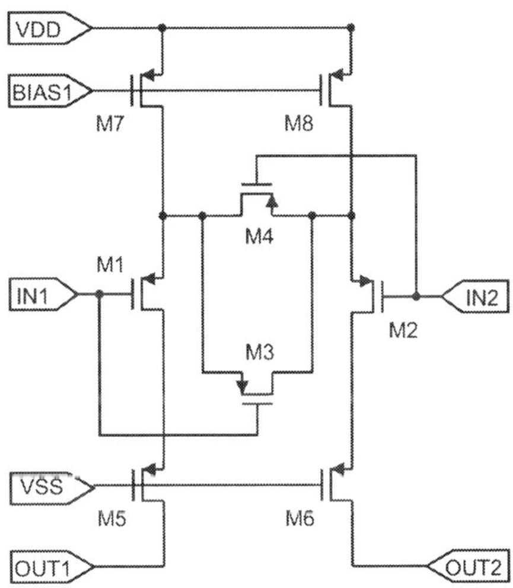

Figure 5-40: *Feedforward coefficient implementation.*

The circuit implementation of the single-bit quantizer is shown in Figure 5-41. The output of the five feedforward coefficients is connected to the drain of M11-M12 and the differential current input is converted into a voltage by the latch M1-M2. The cascode transistors M3-M4 increase the latching gain. The latch is biased by the current sources

M11-M12 with cascodes M9-M10, while the diodes M7-M8 limit the output swing, ensuring a short recovery time. The latch differential output is then fed into AND gates controlled by a delayed version of the quantizer clock and the true digital output is stored in a flip-flop. After a new digital output is available, the analog latch output is reset to half the supply V_{DD} by the switches M5-M6. In this way, the quantizer becomes insensitive to any kind of memory effect. The quantizer phase-shift during the reset phase should be accounted in total loop filter phase-shift in order to guarantee the ΣΔ modulator stability. The errors introduced by the quantizer are not critical because they are suppressed by the 5th-order loop filter. All quantizer control signals are provided by an internal DLL. To ensure the proper operation of the ΣΔ modulator, there is a delay ($\Delta t<20\%$ T_S) between the quantizer clock and the feedback DAC clock.

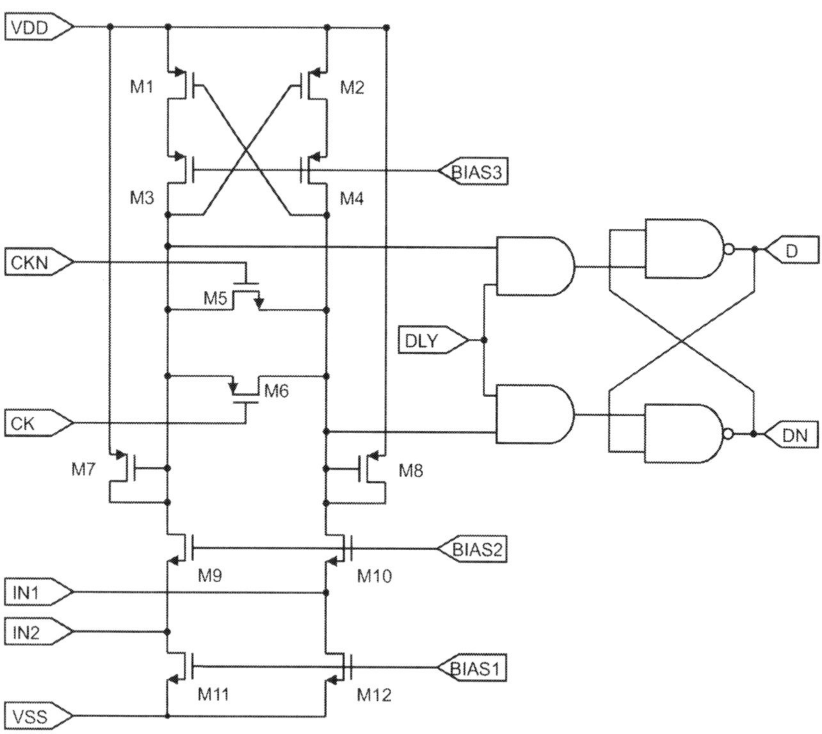

Figure 5-41: *Single-bit quantizer and RS-latch.*

5.6 SC Feedback DAC

The effect of clock jitter in CT ΣΔ modulators was reviewed in section 3.4. Based on the random jitter model described in section 3.4.1, a SC DAC implementation was chosen for this design in order to overcome the *SNR* limitations caused by timing uncertainties in RTZ DACs. According to this model, the *SNR* improvement factor from a RTZ to SC DAC is given by [7]:

$$\Delta SNR_{RTZ \to SC} = 10\log\left[\left(\frac{1-e^{\delta \cdot \alpha}}{\delta \cdot \alpha}\right)^2\right] \tag{5-43}$$

For a 50% DAC duty cycle δ and a relative settling time α equal to 10 time constants, a theoretical *ΔSNR* improvement of 29.4 dB is achieved. Figure 5-42 shows the peak *SNR* limitation as a function of the RMS jitter σ_j, due to pulse-width jitter in RTZ and SC DACs [7], and due to pulse-position jitter [5]. The peak *SNR* due to thermal noise in FM mode is also shown as a reference. Figure 5-42 reveals that an oscillator with σ_j=1ps is required to achieve the target peak *SNR*=98 dB if a RTZ DAC is used in this CT ΣΔ modulator. The SC DAC allows the use of an oscillator with σ_j=5ps, as the pulse-position limitation (independent of α) becomes dominant over the SC DAC pulse-width jitter limitation. The effective *ΔSNR* improvement then becomes 14 dB.

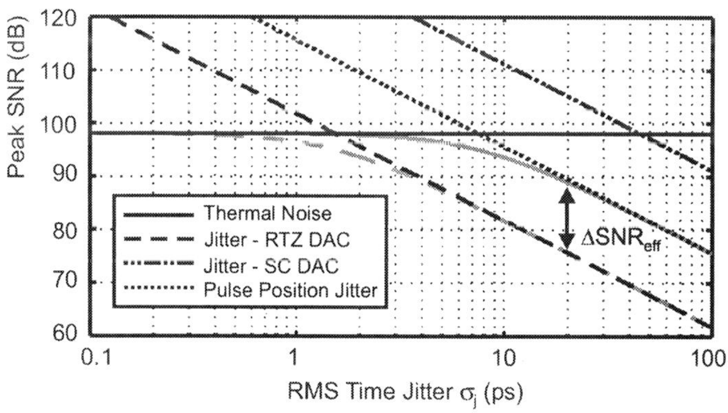

Figure 5-42: *Peak SNR limitation due to random clock jitter for RTZ and SC DACs (FM mode), α=10.*

The combination of (5-25) with the expression for the equivalent resistor of a SC network ($R_{dac}=T_S/C_d$) gives the relation between the DAC capacitor C_d and the other CT ΣΔ modulator parameters:

$$C_d = T_S \cdot \frac{V_{in}}{V_{dac}} \cdot \frac{1}{R_{in}(1-\delta)} \tag{5-44}$$

where V_{in} is the maximum input rms voltage and V_{dac} is the DAC differential voltage. The relation between V_{dac} and the DAC reference voltage V_{ref}=1.6V depends on the actual SC DAC configuration. For the DAC configuration shown in Figure 5-43a, $V_{dac}=V_{ref}$ and C_d=9pF is required. If instead the DAC capacitors are cross-connected to ground and V_{ref} during the discharging phase S2, as depicted in Figure 5-43b, then $V_{dac}=2V_{ref}$ and the required capacitor size becomes C_d=4.5pF.

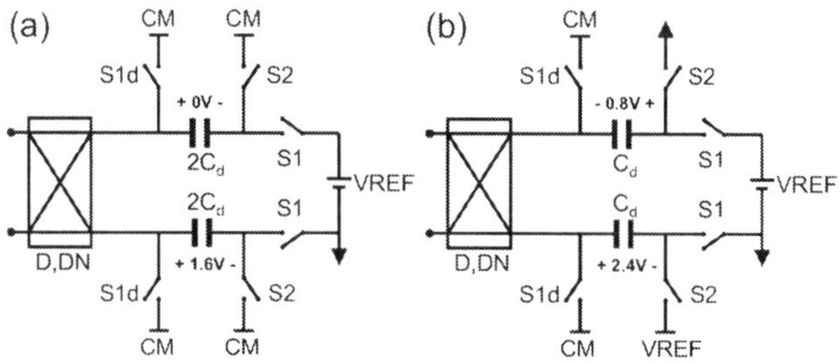

Figure 5-43: *Single-bit SC DAC configurations.*

The complete single-bit SC DAC circuit is shown in Figure 5-44 together with the ΣΔ modulator front-end circuitry. It is implemented with series-connected NMOS-on-nwell floating MOSCAPs (Figure 5-23) and MOS switches [7]. Different switches are implemented with NMOS or PMOS transistors according to the internal DAC voltages. As a result, the discharging transients in the PMOS and NMOS DAC paths are slightly different. Because a single-bit differential DAC is used, the mismatch between the paths only causes an offset error. In the 1st clock phase (Figure 5-45), the capacitors are charged to the common-mode voltage $CM=V_{ref}/2$ by closing switches S1/S1d (switches S2 open). In the 2nd clock phase, switches S2 are closed (switches S1/S1d open) and the capacitors C_d are discharged in a bitstream-dependent manner by closing switches D or DN. As the input of the IF-to-baseband ΣΔ ADC is capacitively coupled to the radio front-end, the input common-mode level of the 1st OTA follows the DAC common-mode voltage *CM*. The DAC output current is subtracted from the input current at the virtual-ground node and integrated on the capacitors of the 1st integrator. To reduce charge injection, S1 opens slightly before S1d, and S2 opens slightly before D or DN. The noise contribution of the SC DAC is calculated in Appendix B.

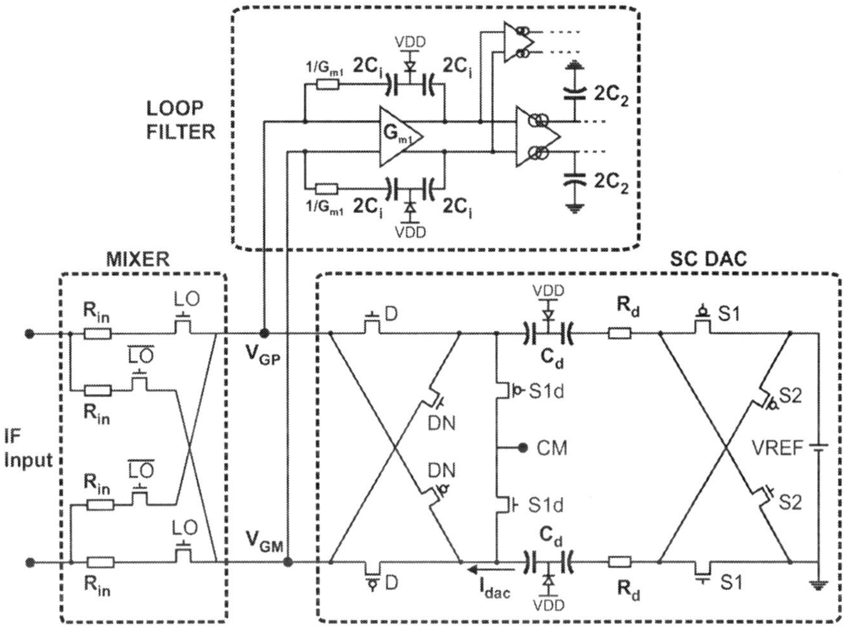

Figure 5-44: *IF-to-baseband ADC input stage: IF mixer, 1st integrator and single-bit switched-capacitor (SC) feedback DAC.*

The total DAC series resistance $R_d' = R_d + 2r_{on}$ is sized in order to allow the DAC output current to settle within 10 time constants $\tau = C_d R_d'$. The poly resistors $R_d = 85\,\Omega$, placed between the capacitors C_d and the bitstream independent switches S1/S2 [6], are used to linearize the settling behaviour of the DAC output current and to limit its peak value. With $R_d' = 265\,\Omega$, I_{dac} is limited to 6.1 mA, avoiding excessive peaking across the virtual-ground nodes V_{GP} and V_{GM} (section 5.3.2), and preventing slewing of the 1st OTA.

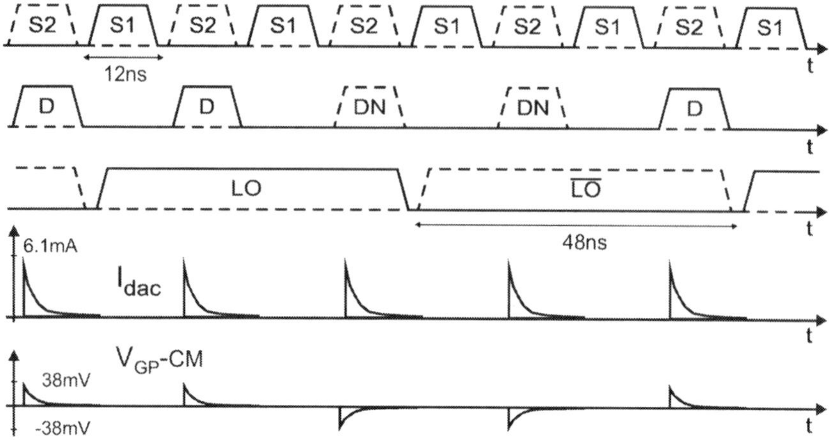

Figure 5-45: *SC feedback DAC and mixer timing diagram, with transient behaviour of I_{dac} and V_{GP} (PSTAR simulation of the 2nd prototype).*

Figure 5-45 shows the timing diagram of the SC DAC and the input passive mixer circuits (2nd prototype), together with the transient behaviour of the DAC output current I_{dac} and V_{GP}. The peak virtual-ground voltage is limited to 38 mV (76 mV differential). As discussed in section 5.2, the passive mixers also require good virtual-ground nodes to implement a linear V-I conversion. To provide a quiet virtual-ground during the mixer transitions, the mixer switches change state only after the 1st OTA input nodes have had enough time to settle (10 time constants) from the SC DAC discharge.

The DAC reference voltage V_{ref} is generated on-chip from an external V_{gap}=1.2 V stable supply. Figure 5-46 shows the reference buffer circuit (2nd prototype) and PSTAR simulation results of the output voltages V_{ref} and *CM*. The reference buffer consists of the input differential pair M1-M7, the voltage follower M8 and a resistor divider. They act as control loop that copies the voltage V_{gap} to node B. As a result, $V_{ref}=4V_{gap}/3$ and $CM=2V_{gap}/3$ in steady-state conditions. Frequency compensation and noise filtering was provided by the capacitor C_C=250pF, and the area of the resistors R=100 Ω was minimized in order to reduce the parasitic capacitance from nodes *A*, *B*, *C* and *D* to ground.

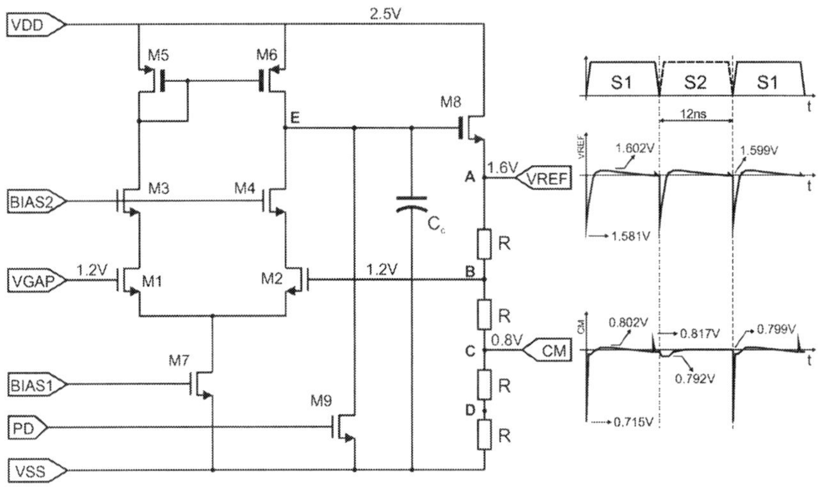

Figure 5-46: *DAC reference buffer and simulation result.*

Every $T_S/2$ interval, one of the DAC capacitors is connected to the V_{ref} terminal and the follower transistor provides the 6.1 mA DAC peak currents at a rate $2f_S$. The control system was designed to allow the voltages V_{ref} and CM to settle within 10 time constants to their final values (Figure 5-46), guaranteeing almost complete charge and discharge of the capacitors C_d. In order to obtain the settling behaviour of the buffer output voltages and currents dominated by the $C_d\,R_d'$ time constant, the buffer transconductance $G_M=A_I \cdot g_{m8}$, where A_I is the voltage gain of the input pair, was designed for $G_M>>1/R_d'$. The bias current for the differential pair is I_b=6mA and the quiescent current for M8 is I_q=4mA. For the proper biasing of transistors M5, M6 and M8, a higher supply voltage (2.5–3.3 V) was used in this circuit. For that reason, these transistors were implemented as thick oxide devices. In order to obtain the DAC noise contribution determined by the $k\theta/C_d$ factor, the noise level at the output node VREF was designed to be non-dominant.

5.7 Experimental Results

Two prototypes of the IF-to-baseband ADC for car radios were fabricated. Both ICs were designed for the specifications described in Table 5-1. The first prototype [8] achieves a DR of 110 dB in AM mode, and 91 dB in FM mode. The second prototype [1] achieves the target DR of 118dB in AM mode and 98dB in FM mode, thanks to the redesigns of the 1st integrator, the feedback SC DAC and the DAC reference buffer. Special attention was also paid to the timing between the DAC and the mixer, and to the overall physical design and lay-out.

5.7.1 First Prototype

The prototype chip shown in Figure 5-47 was fabricated in a 1P 5M 0.18μm digital CMOS process. The IC includes a clock divider, a DLL, two DAC reference (V_{ref}) buffers, and LVDS transmitters for the I/Q bitstream digital outputs. The DLL generates delayed clocks to control the quantizer and the V_{ref} buffers provide the SC DACs with an 1.6V reference generated from a stable 1.2V external supply. The 41.7MHz clock was provided externally. The clock divider generates two I/Q mixer clocks at 10.425MHz. The modulator operates from a 1.8V supply and consumes 190mW (I+Q). The active area is 3.6 mm^2.

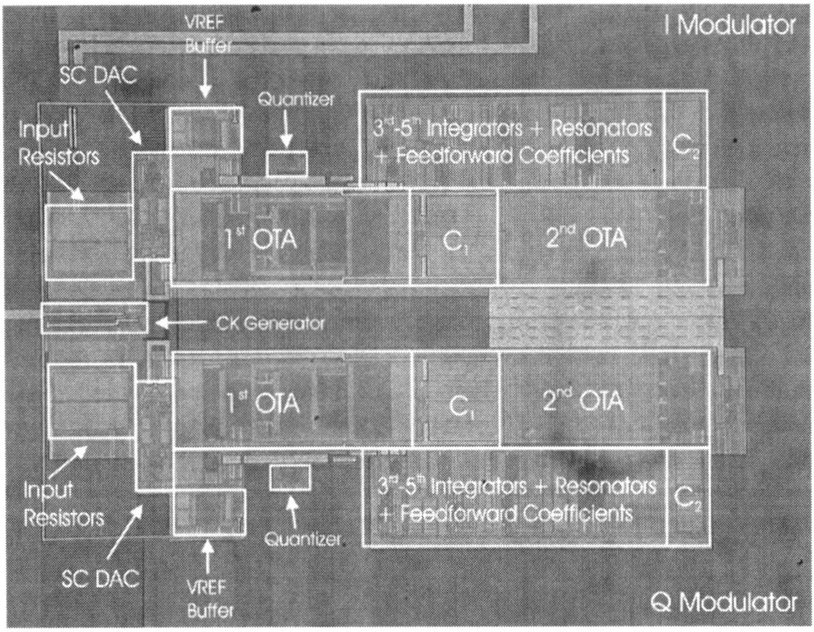

Figure 5-47: *First prototype chip micrograph.*

The ΣΔ modulator' s *SNDR* is plotted as a function of the input signal in Figure 5-48. In AM mode (3kHz BW) the DR is 110 dB and peak *SNR* is 103dB. In FM mode (200 kHz BW) the DR is 90 dB and the peak *SNR* is 83dB. The full scale (FS) input is 0.5 Vrms (-6dB). The achieved DR is 10dB better than previous designs [2], [3], but it is still 8dB below the target specification (Table 5-1).

Figure 5-49 shows the measured ΣΔ modulator complex output spectrum for a -6dBFS two-tone input signal (10.625 and 10.675 MHz) with the mixer ON. The IM_2 distance is 90 dB and the IM_3 distance is 85dB and the SFDR is greater than 70dB in the IBOC band (25–525kHz). The comparison between this measurement and the simulation presented in Figure 5-7 reveals that the in-band noise floor is above the expected. The measured complex output spectrum is shown again in Figure 5-50 for an idle input. This measurement reveals that the signal band is covered by a higher noise floor and several spurious tones can be seen within the signal band. It is not possible to see the asymmetrical spectrum provided by the complex noise shaping. Due to the higher noise floor, the ΣΔ modulator *SNR* is degraded and the input DR is reduced.

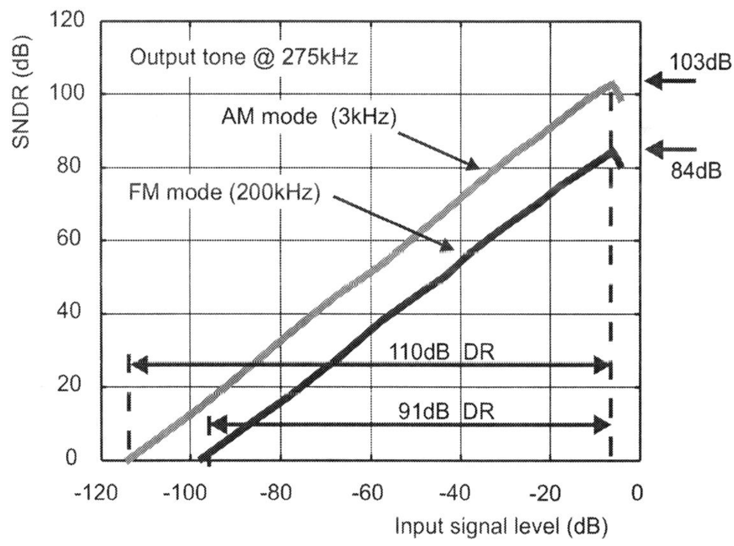

Figure 5-48: *Measured SNDR as function of the input signal in AM and FM modes.*

To understand the reason for the high noise level and the spurious tones inside the signal band, a measurement with a small dc input voltage was performed. For this experiment the f_S was set to be 42MHz. The results are shown in Figure 5-51. A single -12dB tone is located 200kHz below $f_S/2$ (Figure 5-51b). The frequency location of this tone near $f_S/2$ depends on the dc offset applied at the input of the modulator [16]. At low frequencies (Figure 5-51a), a replica of this tone is present and located 400kHz far from dc. The frequency of this tone is exactly twice the 200kHz distance between the high-frequency tone and $f_S/2$.

Based on the analysis presented in section 3.3, the measurements shown in Figures 5-50 and 5-51 lead to the conclusion that the in-band noise power was increased due to the down-conversion of the spectrum (noise floor and tones) near $f_S/2$. There is a factor 2 between the low-frequency tone location, and the distance between $f_S/2$ and the original tone. This indicates that the down-conversion of the high-frequency spectrum is caused either by 2nd-order harmonic distortion from the loop filter building blocks, either by bitstream dependent loading of the DAC reference voltage, or by both mechanisms together.

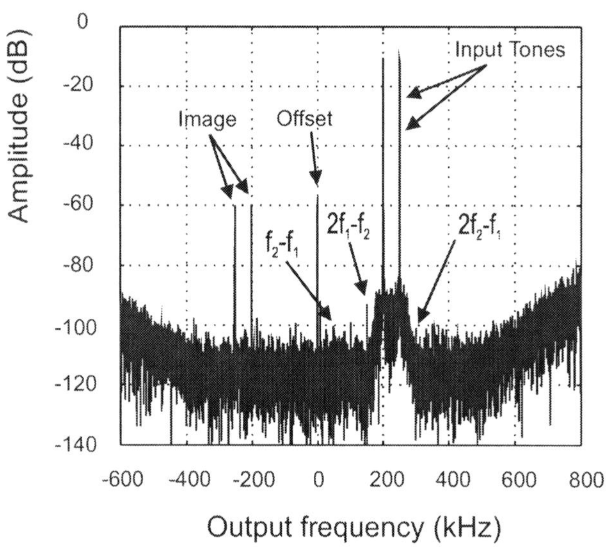

Figure 5-49: *Meas. complex output spectrum for a two-tone input (10.625 and 10.675 MHz) and RBW=100Hz.*

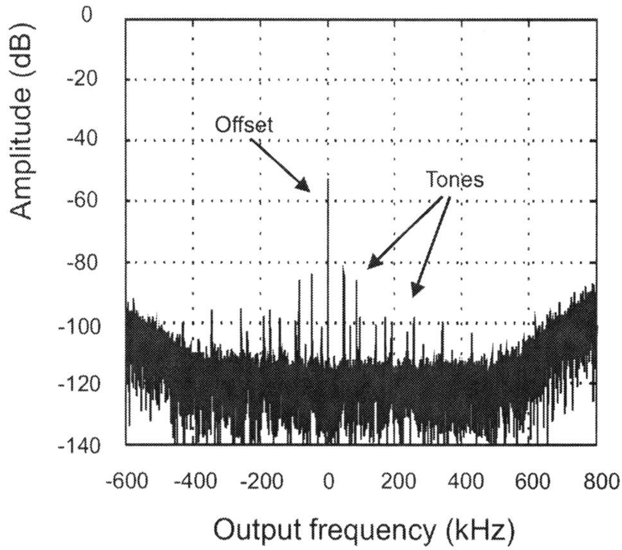

Figure 5-50: *Measured complex output spectrum for an idle input (RBW=100Hz).*

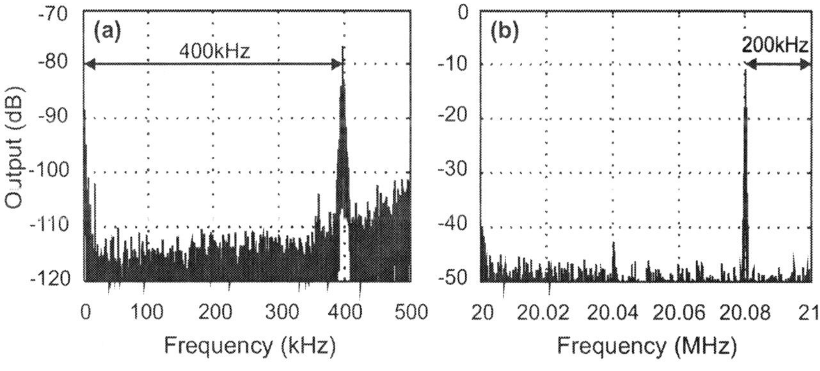

Figure 5-51: *Measured output spectrum for a dc input. Zoom near the dc (a), zoom near $f_S/2$ (b).*

The following paragraphs describe the measures adopted to improve the performance of the first prototype ADC.

First prototype analysis and improvements

The analysis of the design and the lay-out of the first prototype revealed several faults. They are listed below:

– The parasitic resistance of the laid-out power lines (conveying almost 100mA) was large enough to reduce the effective supply voltage available for the 1st and 2nd integrators to almost 1.6V. The bias generators for the loop filter OTAs were connected to a ground line different from the high-ohmic ground line used for the 1st and 2nd OTAs and voltage bias distribution was used instead of current bias distribution. As a result, the operating point of the 1st and 2nd OTAs changed considerably causing an increase of the non-linearity in the first fabricated IC. Because of the internal difference of potential between the power lines supplying the OTAs and the biasing generators, even an increase in the nominal supply from 1.8V to 2.2V could not improve the performance of the ADC under test. Furthermore, the input CM voltage, set by the average of the feedback bitstream, could not be increased significantly in order to cancel the voltage difference between the different ground lines. The output spectrum (complete circuit PSTAR simulation) of the ΣΔ modulator is shown in Figure 5-52 for R_{VDD}=R_{VSS}=0.1, 0.5, 1, 2, 3Ω and VDD=2V. The simulated model accounts for the parasitic series resistance in all different laid-out ground lines. R_{VDD} and R_{VSS} are the resistors in series with the power lines supplying the 1st and 2nd OTAs. Figure 5-52 shows that for R_{VDD} and R_{VSS} larger than 0.5Ω, it is possible to see a significant increase in the quantization noise floor within the positive-sided signal band.

– The large triode-biased PMOS transistors used for the common-mode control of the 1st OTA (Figure 5-32) in the first prototype chip increased considerably the load capacitance at 1st OTA output node, reducing the unity gain frequency ω_u. As described in section 5.3.2, the peaking across the virtual-ground node is inversely proportional to ω_u.

– The floating capacitors of the 1st integrator are series-compensated MOSCAPs. The four MOSCAPs required to implement the two integrating capacitors were laid-out with poor matching, what compromised the cancellation of the 2nd-order non-linearity at the 1st integrator, as described in Appendix A.

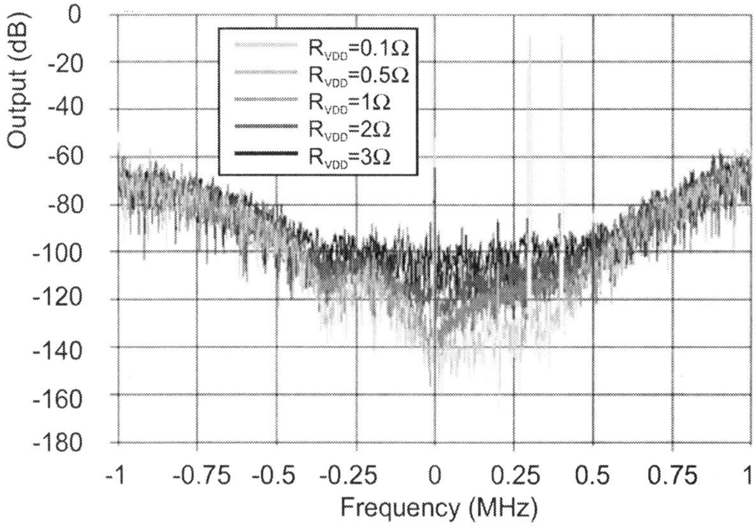

Figure 5-52: *Complex output spectrum (PSTAR simulation) for VDD=2.0 V, and parasitic resistors* R_{VDD}= R_{VSS}*=0.1, 0.5, 1, 2, 3 Ω*

– The SC DAC and the DAC V_{ref} buffer were connected to a ground line different from the ground line used for the 1st integrator. This ground line was also characterised by high parasitic resistance and a ground loop.

– The DAC V_{ref} buffer was designed with very low bias current and, consequently, with high output thermal noise. Furthermore, a pin for the connection of an external decoupling capacitor was not available in the 1st fabricated IC. It was verified by measurements and simulation that a significant part of the high in-band noise level (Figures 5-49 and 5-50) was caused by the noisy DAC reference voltage.

– Inappropriate mixer timing: the transition between the mixer phases LO and $\overline{LO}$ (and vice-versa) was synchronized with the rising edge of DAC phase S2. In this way, the mixer changed state when a large voltage was present at the virtual-ground nodes of the 1st integrator.

Based on the analysis of the 1st prototype, some improvements were implemented during the design and lay-out of the 2nd prototype:

– Lay-out of very low ohmic power lines and use of single-point star topology for VDD and ground distribution. The 2nd prototype IC is free of ground/VDD loops.

– Use of current bias distribution for the 1st and 2nd integrators. The bias generators were also connected to the same power lines as the integrators.

– Redesign of the 1st OTA, using a common-mode control scheme based on an error amplifier (Figure 5-32). As a result, the OTA unity gain frequency was increased from 380 to 610MHz. The voltage drop across the triode biased transistor was also eliminated.

– The series-compensated MOSCAPs used as floating capacitors in the 1st integrator and the 2nd integrator grounded MOSCAPs, were laid-out in a common-centroid configuration (shown in Figure 5-53b for the 1st integrator) to improve matching and hence the non-linearity cancellation. The unity capacitors were also laid-out with small *L*, in order to minimize the parasitic series resistance.

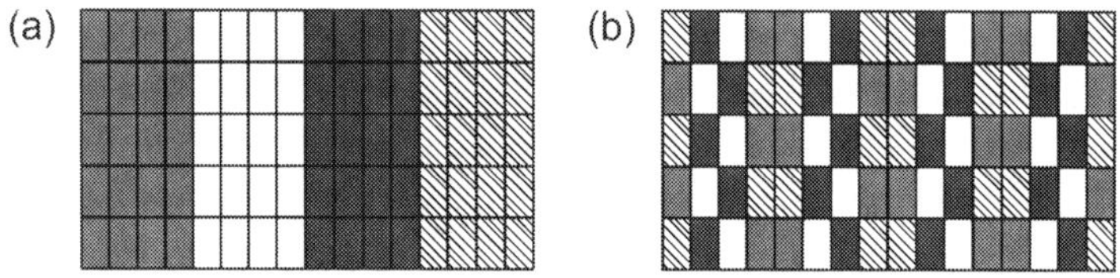

Figure 5-53: *Lay-out of the 1st integrator capacitor array, (a) 1st prototype, (b) 2nd prototype.*

– The SC feedback DAC internal logic was changed in such way that the transition between the mixer phases LO and $\overline{\text{LO}}$ was synchronized with the rising edge of DAC phase S1, as depicted in Figure 5-45.

– The feedback DAC reference buffer was redesigned for lower thermal noise and a large external decoupling capacitor was used to limit the noise bandwidth. Figure 5-46 shows the reference buffer designed for the 2nd prototype.

The combined effect of all these improvements is assumed to be responsible for the increased DR and linearity of the 2nd prototype IC.

5.7.2 Second Prototype

Figure 5-54 shows the micrograph of the second prototype chip. It was fabricated in a 1P 5M 0.18 μm digital CMOS process. Similar to the first prototype, this IC contains a clock divider, a DLL, two DAC reference (V_{ref}) buffers, and LVDS transmitters. The DLL generates delayed clocks to control the quantizer and the V_{ref} buffers generate 1.6V from a stable external 1.2V reference. A pair of CT feedback DACs was also integrated with this prototype for comparison purposes. However, due to an error in the scaling of the DAC resistors, the measurements obtained were not useful. The 41.7MHz clock is provided externally. The modulator operates from a 1.8V supply and consumes 210mW (I+Q). The total active area is 6.0 mm^2. The area of each ΣΔ modulator is 2.1 mm^2.

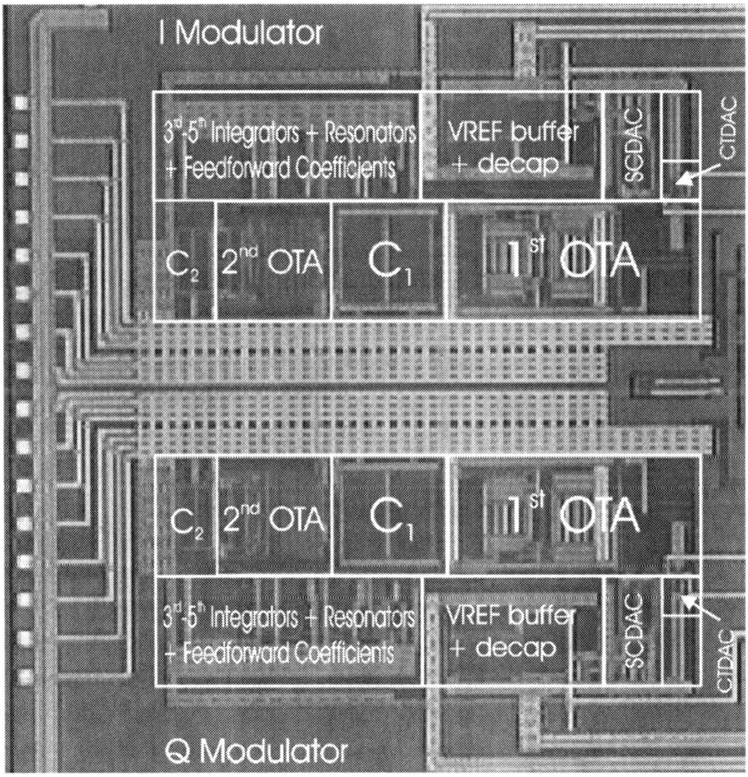

Figure 5-54: *Second prototype chip micrograph.*

Figure 5-55 shows the ΣΔ modulator's *SNDR* as a function of the input signal in AM and FM modes. The input tone is at 10.7 MHz and it is down-converted to the 275 kHz low-IF. Again, the full scale (FS) input is 0.5 Vrms (-6 dB). In AM mode (3 kHz BW) the DR is 118 dB and peak *SNR* is 113 dB. In FM mode (200 kHz BW) the DR is 98 dB and the peak *SNR* is 90 dB. This IF-to-baseband ΣΔ ADC prototype achieves the DR specified in Table 5-1. In IBOC mode (not shown in this picture), the DR is 86 dB and the peak *SNDR* is 78 dB.

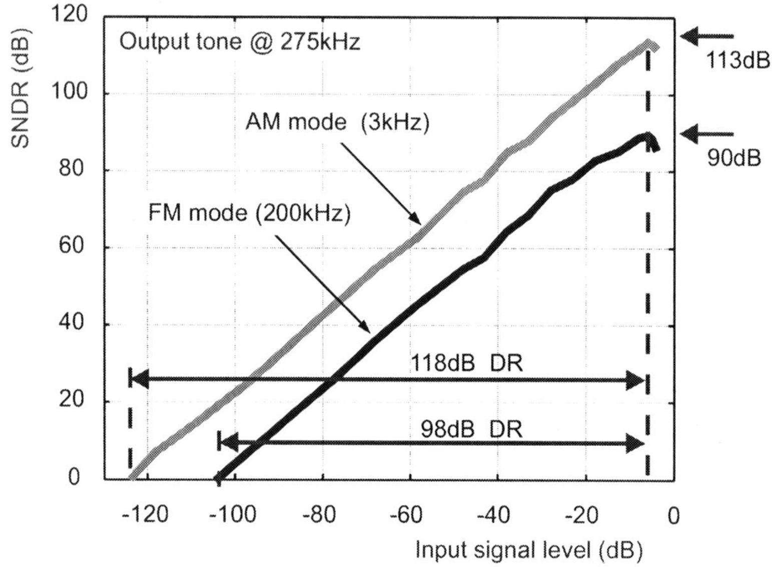

Figure 5-55: *Measured SNDR as a function of the input level in AM and FM modes.*

The measured ΣΔ modulator real (mixer OFF) output spectrum is shown in Figure 5-56. The input is a -6dBFS two-tone signal at 250 and 300kHz. The modulator is very linear with the mixer OFF: the IM_2 distance is 102dB, and the IM_3 distance is 98dB. The SFDR is greater than 90dB within the IBOC bandwidth (25–525kHz). Figure 5-57 shows the measured ΣΔ modulator complex (mixer ON) output spectrum. The input is a -6dBFS two-tone signal at 10.625 and 10.675 MHz. The absolute input amplitude is 4dB higher than in the previous measurement to compensate for the passive mixer -4dB gain. In this case the IM_2 distance is 92dB, and the IM_3 distance is 91dB. The SFDR is 88dB in the 500kHz IBOC band. Some spurious tones can be seen inside the signal band. The strongest tones are in fact coming from the arbitrary waveform generator (AWG). The signal generator used in this measurement is the same as used in Figure 5-56. For the measurements with the mixer ON however, the AWG oversampling ratio is reduced by a factor 40 and its output noise is increased. The high noise floor around the two input tones is due to the multiplication of the mixer clock phase-noise with the AWG phase-noise. Because of the noise coming from the signal generator, it is not possible to clearly see the asymmetrical noise shaping.

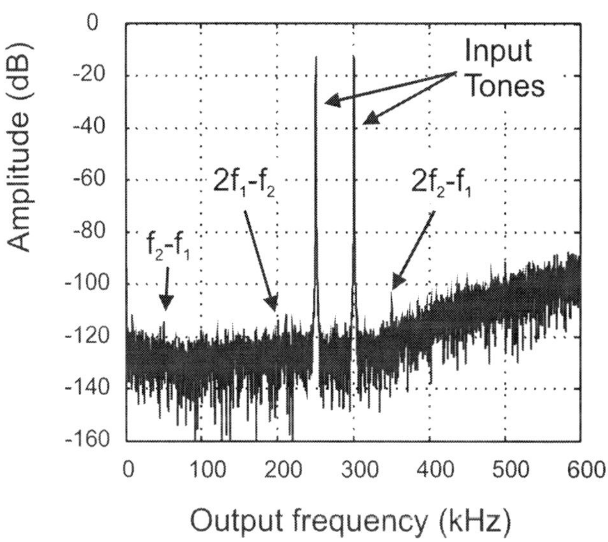

Figure 5-56: *Measured real output spectrum. Two-tone test (250 and 300kHz) with mixer OFF and RBW=100Hz.*

The complex output spectrum for an idle input measurement is shown in Figure 5-58 (black plot). In this case the measured spectrum is very similar to the simulated spectrum shown in Figure 5-7 and it is possible to see the asymmetrical noise shaping. The left side of the spectrum (image band) is quantization-noise-limited, while the right side (signal band centred on +275kHz) is thermal-noise-limited. Also depicted in

Figure 5-58 is the measured output spectrum for an idle input obtained with the first prototype (gray plot). The comparison between both plots reveals how much the noise floor was reduced due to the improvements applied to the design of the second prototype. Also, the power of the in-band down-converted spurious tones was reduced by at least 30dB compared with the first prototype.

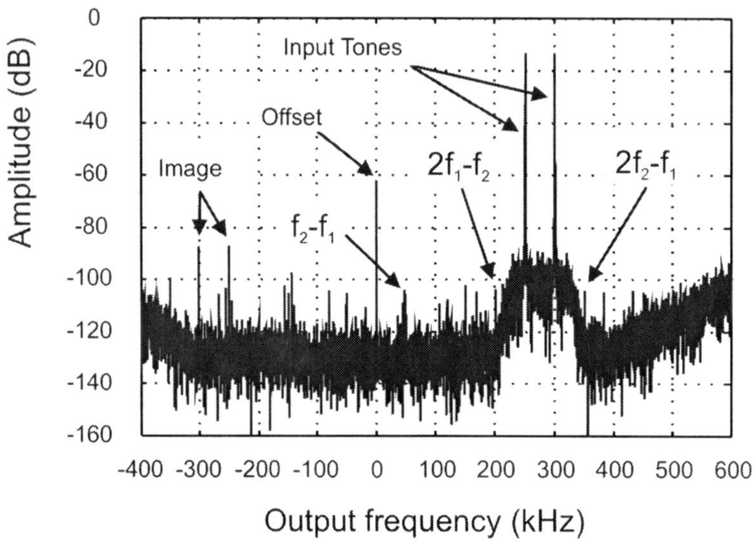

Figure 5-57: *Measured complex output spectrum. Two-tone test (10.625 and 10.675MHz) with mixer ON and RBW=100Hz.*

The image tones are 70dB below the input tones for the measurement shown in Figure 5-57. The image rejection ratio (*IRR*) was measured in all 32 available samples. An *IRR* histogram based on these measurements (mixer ON) is presented in Figure 5-59. The mean *IRR* is 63dB and the standard deviation is $\sigma = 6$dB. The 3σ lower boundary for the *IRR* is 45dB, 5dB beyond the 40dB *IRR* required exclusively from the quadrature IF-to-baseband ADC (Table 5-1). This result was obtained exclusively with prudent lay-out of the ADC building blocks. Dynamic element matching techniques were not used in this design. Table 5-3 shows the summary of performance for the second prototype IF-to-baseband ADC.

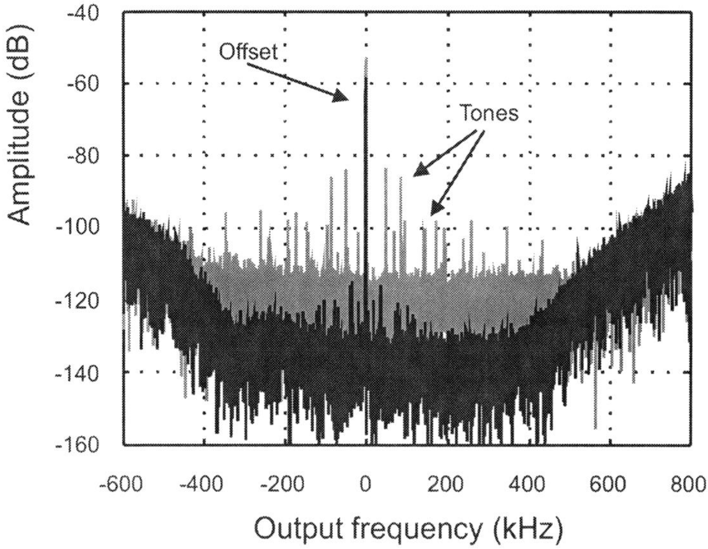

Figure 5-58: *Measured complex output spectrum for an idle input (RBW=100Hz) obtained with the 1st prototype (gray plot) and with the 2nd prototype (black plot).*

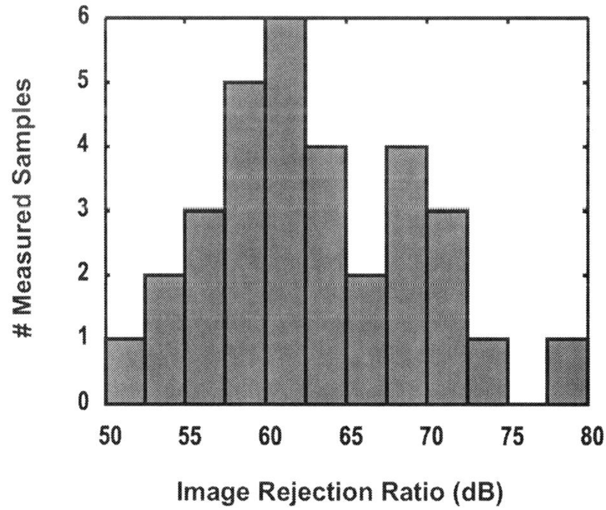

Figure 5-59: *Image rejection ratio (IRR) histogram.*

Table 5-3. *Summary of performance for the 2nd prototype.*

Process	1P, 5M 0.18 μm digital CMOS		
Supply voltage	1.8 V		
ΣΔ modulator	5th-order complex CT, 1b SC DAC		
Max input voltage	0.5 V_{rms}, differential		
Sampling rate	41.7 MHz		
Mode	AM	FM	IBOC
Signal bandwidth	3 kHz	200 kHz	500 kHz
Dynamic range	118 dB	98 dB	86 dB
Peak SNDR	113 dB	90 dB	78 dB
SFDR	88 dB		
Intermodulation	Mixer Off	Mixer On	–
IM_2	102 dB	92 dB	–
IM_3	98 dB	91 dB	–
IRR	typical 63 dB, σ= 6 dB		
Active area	6.0 mm^2		
Power consumption	210 mW (I+Q)		

References

1. P. Silva, L. Breems, K. Makinwa, R. Roovers and J. Huijsing, "An IF-to-Baseband ΣΔ Modulator for AM/FM/IBOC Radio Receivers with a 118 dB Dynamic Range", *IEEE J. Solid-State Circ.*, vol. 42, 1076–1089, May 2007.
2. Q. Sandifort, L.J. Breems, C. Dijkmans and H. Schuurmans, "IF-to-Digital Converter for FM/AM/IBOC Radio", Proceedings of the 29th European Solid-State Circuit Conference, Estoril, Portugal, Sept. 2003, pp. 707–710.
3. E.J. van der Zwan, K. Philips and C.A.A. Bastiaansen, "A 10.7-MHz IF-to-Baseband Sigma Delta A/D Conversion System for AM/FM Radio Receivers", *IEEE J. Solid-State Circ.*, vol. 35, 1810–1819, Dec. 2000.

4. L.J. Breems, E.J. van der Zwan, E.C. Dijkmans and J.H. Huijsing, "A 1.8mW CMOS ΣΔ Modulator with Integrated Mixer for A/D Conversion of IF Signal", *IEEE J. Solid-State Circ.*, vol. 35, 468–475, Apr. 2000.
5. E.J. van der Zwan and E.C. Dijkmans, "A 0.2-mW CMOS ΣΔ Modulator for Speech Coding with 80-dB Dynamic Range", *IEEE J. Solid-State Circ.*, vol. 31, 1873–1880, Dec. 1996.
6. M. Ortmanns, F. Gerfers and Y. Manoli, "A Continuous-Time ΣΔ Modulator with Reduced Sensitivity to Clock Jitter Through SCR Feedback", *IEEE Trans. CAS-I.*, vol. 52, 875–884, May 2005.
7. R.H.M. van Veldhoven, "A Triple-Mode Continuous-Time ΣΔ Modulator with Switched-Capacitor Feedback DAC for a GSM-EDGE/CDMA2000/UMTS Receiver", *IEEE J. Solid-State Circ.*, vol. 38, 2069–2076, Dec. 2003.
8. P. Silva, L. Breems, K. Makinwa and J. Huijsing, "A 110dB Dynamic Range Continuous-Time IF-to-Baseband Sigma-Delta Modulator for AM/FM/IBOC Receivers", Proceedings of the 2006 ISCAS, Kos, Greece, May 2006.
9. L. Breems and J.H. Huijsing, "Continuous-Time Sigma-Delta Modulation for A/D Conversion in Radio Receivers", *Kluwer Academic, Boston, MA,* 2001.
10. B. Razavi, "Design of Analog CMOS Integrated Circuits", *McGraw-Hill*, New York, 2001
11. H. Yoshizawa, Y. Huang, P.F. Ferguson Jr. and G.C. Temes, "MOSFET-Only Switched-Capacitor Circuits in Digital CMOS Technology", *IEEE J. Solid-State Circ.*, vol. 34, 734–747, June 1999.
12. A.T. Behr, M.C. Schneider, S. Noceti Filho and C.G. Montoro, "Harmonic Distortion Caused by Capacitors Implemented with MOSFET Gates", *IEEE J. Solid-State Circ.*, vol. 27, 1470–1475, Oct. 1992.
13. J.H. Huising, "Operational Amplifiers: Theory and Design", *Kluwer Academic, Bostan, MA*, 2001.
14. K. Bult and G.J.G.M. Geelen, "A Fast-Settling CMOS Op Amp for SC Circuits with 90-dB DC Gain", *IEEE J. Solid-State Circ.*, vol. 25, 1379–1384, Dec. 1990.
15. F. Krummenacher and N. Joehl, "A 4-MHz CMOS Continuous-Time Filter with On-Chip Automatic Tuning", *IEEE J. Solid-State Circ.*, vol. 23, 750–758, June. 1983.
16. S.R. Norsworthy, R. Schreier and G.C. Temes, "Delta-Sigma Data Converters: Theory, Design and Simulation", *IEEE Press, Piscataway, NJ*, 1997.

Conclusion 6

This chapter presents the final conclusions of this research project. The main features and the benchmarking of the prototype ADC are presented in section 6.1. The economic feasibility of the commercial use of the proposed car radio receiver architecture is then discussed in section 6.2.

6.1 Benchmarking

The design of a high-resolution IF-to-baseband ΣΔ ADC for AM/FM/IBOC receivers was the goal of this research. The designed ΣΔ quadrature ADC combines the anti-aliasing suppression of a CT loop filter, the low jitter sensitivity of a SC feedback DAC, and the low power consumption of passive mixers for low-IF down-conversion. The achieved 118dB DR in AM mode (3kHz BW) allows the realization of a car radio with a reduced number of external components. The AM ceramic channel selection filter and low-noise/high-linearity VGA required in the traditional AM/FM DSP based receiver architecture are no longer necessary. As a result, the radio's noise figure and its sensitivity in AM mode are improved. In the proposed radio receiver, most of the AM channel selection is performed in the digital domain.

Several high-resolution ΣΔ modulators have been published in the scientific literature since the end of the 1990s. In order to better compare the performance of the prototype ADC [1] with the state-of-the-art, Table 6-1 lists all ΣΔ ADCs with a DR greater than 100dB in the 3kHz BW presented at the IEEE International Solid-State Circuits Conference (ISSCC) since 2000.

Table 6-1. *Comparison of high-resolution ΣΔ ADCs.*

Ref.	Max. input (V_{rms})	BW (kHz)	DR (dB) in 3kHz	Power (mW)	Noise ($nV/\sqrt{Hz}$)
[1]	0.5	200	118	210	11.5
[2]	0.55	20,000	114	20	18.6
[3]	–	8,500	119	375	–
[4]	0.71	2,200	114	14	20.1
[5]	0.55	1,100	110	5.4	28.1
[6]	–	4,000	104	38.2	–
[7]	1.2	1,000	125	362	7.9
[8]	0.71	10,000	110	185	33.5
[9]	2	20	114	18	35.4
[10]	2	20	110	37.3	56.4
[11]	0.42	2,000	102	3	64.6
[12]	–	600	109	5.4	–
[13]	0.07	600	100	12	34.1
[14]	0.55	23,000	111	43	25.9
[15]	0.071	10,000	102	122	26.6
[16]	0.55	12,500	120	200	9.4
[17]	0.35	1,000	102	6	57.1
[18]	0.5	300	108	8	36.4
[19]	0.42	2,500	101	150	72.8
[20]	2	20	122	55	14.1
[21]	1	1,228	109	8.2	45.2
[22]	0.53	1,100	114	62	19.5

(Continued)

Table 6-1. *Comparison of high-resolution ΣΔ ADCs (Continued).*

Ref.	Max. input (V_{rms})	BW (kHz)	DR (dB) in 3kHz	Power (mW)	Noise (nV/√Hz)
[23]	0.1	1,000	101	4.4	35.4
[24]	–	333	110	50	–
[25]	0.9	270	105	24	64.7
[26]	0.84	1,100	114	180	24.6
[27]	0.63	200	102	8	79.2
[28]	–	180	102	5	–
[29]	–	2,000	123	90	–
[30]	1.4	1,250	123	152	10.6
[31]	1.4	1,250	116	105	23.7
[32]	–	2,000	103	100	–

In order to achieve 118dB DR in 3kHz, this ADC was designed with a noise floor 18dB lower than the previous generation of IF-to-baseband ADCs for car radios [33]. This is equivalent to an improvement factor of 8 times in resolution. Table 6-1 reveals that the ADC described in this book achieved the 4th lowest thermal noise floor in comparison to the reported performance of the ΣΔ ADC's presented at the ISSCC since 2000.

6.2 Economic Feasibility

The economic motivations of this project were the elimination of the AM channel selection filter and the VGA from the radio receiver. In production quantities, the first costs US$0.35 per unit, and the second US$0.10 per unit [34]. In order to be economically feasible, the increase in the fabrication costs due to the substitution of the current IF-to-baseband ADC by the high-resolution IF-to-baseband ADC [1] should be below US$0.45 per unit. The area of the IF-to-baseband ADC discussed in [33] is 1.2mm^2 (I+Q). This means that the new ADC is 5 times larger than the previous design. As the fabrication cost of an IC in a 0.18μm CMOS process is about US$0.12 per mm^2, at 2006 prices [35], the extra cost of the new IC is US$0.58 per unit. At this price, part of the economic advantage of the elimination of the filter and VGA is lost.

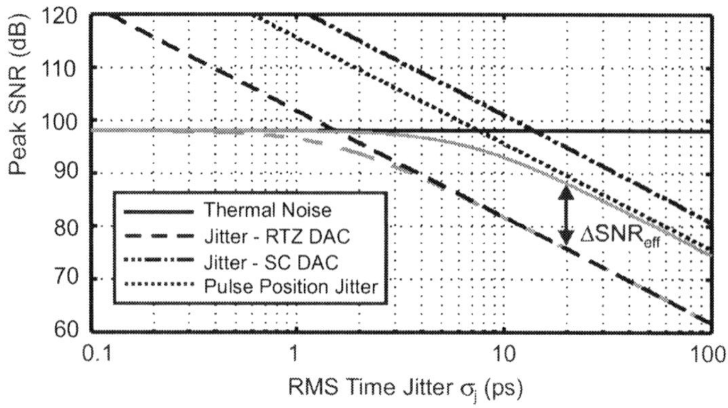

Figure 6-1: *Peak SNR limitation due to random clock jitter for RTZ and SC DACs (FM mode), α=6.5 .*

However, this prototype ADC was over designed regarding the settling time for the SC DAC discharge. As shown in Figure 6-1, a factor α=6.5 could have been used instead of α=10 (Figure 5-42). This is possible because the ultimate jitter-induced SNR limitation for α>5.5 is due to pulse-positioning jitter.

As a result of using α=6.5, the total series resistance of the SC DAC could be increased to 411Ω. This means that the peak current on the feedback path would be reduced from 6.1mA to 3.9mA. According to (5-32), the required 1st OTA transconductance to limit the virtual-ground peaking to 50mV would then become:

$$G_{m1} = \frac{1.6V}{50mV \cdot 411\Omega} = 78mS \qquad (6\text{-}1)$$

In this case, the tail current of the 1st OTA could be scaled down to 26mA and the total ADC power consumption would be reduced to 160mW. The area used by the 1st OTA and the power lines could then be reduced by 1/3. This translates roughly to a 30% area reduction to the whole IC. In this case, the added fabrication cost of the high-resolution IF-to-baseband ADC would be reduced to US$0.36 and the net cost reduction would become US$0.09 per radio. Extra cost reduction can be expected from simplified logistics and assembling, and from the PCB area reduction. As a result, the proposed car radio architecture would then become truly economically attractive.

The main disadvantage of combining an IF-to-baseband CT ΣΔ modulator with a SC feedback DAC, is the increase in power consumption. When the power consumption is dominated by the 1st integrator, as is the case of loop filters with feedforward compensation, the expected power consumption advantage of CT over SC implementations is diminished. This occurs because the 1st OTA also has to be designed to cope with the more stringent transient requirements of a SC integrator.

Suppose instead, a RTZ feedback DAC were employed in this IF ADC design. In this case, G_{m1}=60mS would be enough to obtain the 1st OTA thermal noise below the input resistors thermal noise floor (section 5.3.1). Such OTA could be designed with a tail current I_b=20mA, resulting in an ADC with 132mW total power consumption, that is about 63% of the power consumption of the actual CT IF-to-baseband ΣΔ modulator with SC DAC. The area used by the 1st OTA and the power lines could then be reduced by 1/2, resulting in almost 45% area saving in comparison with the last fabricated prototype IC.

References

1. P. Silva, L. Breems, et al., "A 118dB DR CT IF-to-Baseband ΣΔ Modulator for AM/FM/IBOC Radio Receivers", *ISSCC 2006 Digest of Technical Papers*, Feb. 2006.
2. G. Mitteregger, C. Ebner, et al., "A 14b 20mW 640MHz CMOS CT ΔΣ ADC with 20MHz Signal Bandwidth and 12b ENOB", *ISSCC 2006 Digest of Technical Papers*, Feb. 2006.
3. R. Schreier, N. Abaskharoun, et al., "A 375mW Quadrature Bandpass ΔΣ ADC with 90dB DR and 8.5MHz BW at 44MHz", *ISSCC 2006 Digest of Technical Papers*, Feb. 2006.
4. S. Kwon and F. Maloberti, "A 14mW Multi-bit ΔΣ Modulator with 82dB SNR and 86dB DR for ADSL2+", *ISSCC 2006 Digest of Technical Papers*, Feb. 2006.
5. K-S. Lee, S. Kwon and F. Maloberti, "A 5.4mW 2-Channel Time-Interleaved Multi-bit ΔΣ Modulator with 80dB SNR and 85dB DR for ADSL", *ISSCC 2006 Digest of Technical Papers*, Feb. 2006.

6. Y. Fujimoto, Y. Kanazawa, et al., "A 80/10MS/s 76.3/70.1dB SNDR ΔΣ ADC for Digital TV Receivers", *ISSCC 2006 Digest of Technical Papers*, Feb. 2006.
7. R. Brewer, J. Gorbold, et al., "A 100dB SNR 2.5MS/s Output Data Rate ΔΣ ADC", *ISSCC 2005 Digest of Technical Papers*, Feb. 2005.
8. A. Bosi, A. Panigada, G. Censura and R. Castello, "An 80MHz 4x Oversampled Cascaded ΔΣ-Pipelined ADC with 75dB DR and 87dB SFDR", *ISSCC 2005 Digest of Technical Papers*, Feb. 2005.
9. K. Nguyen, B. Adams, et al., "A 106dB SNR Hybrid Oversampling ADC for Digital Audio", *ISSCC 2005 Digest of Technical Papers*, Feb. 2005.
10. P. Morrow, M. Chamarro, et al., "A 0.18 μm 102dB-SNR mixed CT SC audio-band ΣΔ ADC", *ISSCC 2005 Digest of Technical Papers*, Feb. 2005.
11. L. Dörrer, F. Kuttner, P. Greco and S. Derksen, "A 3mW 74dB SNR 2MHz CT ΣΔ ADC with a tracking-ADC-Quantizer in 0.13μm CMOS", *ISSCC 2005 Digest of Technical Papers*, Feb. 2005.
12. A. Das, R. Hezar, et al., "A 4th-order 86dB CT ΔΣ ADC with two amplifiers in 90nm CMOS", *ISSCC 2005 Digest of Technical Papers*, Feb. 2005.
13. P. Fontaine, A. Mohieldin and A. Bellaour, "A Low-Noise Low-Voltage CT ΣΔ Modulator with Digital Compensation of Excess Loop Delay", *ISSCC 2005 Digest of Technical Papers*, Feb. 2005.
14. N. Yaghini and D. Johns, "A 43mW CT Complex ΣΔ ADC with 23MHz of Signal Bandwidth and 68.8dB SNDR", *ISSCC 2005 Digest of Technical Papers*, Feb. 2005.
15. L. Breems, "A Cascaded Continuous-Time ΣΔ Modulator with 67dB Dynamic Range in 10MHz Bandwidth", *ISSCC 2004 Digest of Technical Papers*, Feb. 2004.
16. P. Balmelli and Q. Huang, "A 25 MS/s 14 b 200 mW ΣΔ Modulator in 0.18 μm CMOS", *ISSCC 2004 Digest of Technical Papers*, Feb. 2004.
17. B. Putter, "ΣΔ ADC with Finite Impulse Response Feedback DAC", *ISSCC 2004 Digest of Technical Papers*, Feb. 2004.
18. R. Gaggl, M. Inversi and A. Wiesbauer, "A Power Optimized 14-bit SC ΔΣ Modulator for ADSL CO Applications", *ISSCC 2004 Digest of Technical Papers*, Feb. 2004.

19. F. Ying and F. Maloberti, "A Mirror Image Free Two-Path Bandpass ΣΔ Modulator with 72 dB SNR and 86 dB SFDR", *ISSCC 2004 Digest of Technical Papers*, Feb. 2004.

20. Y. Yang, A. Chokhawala, et al., "A 114 dB 68 mW Chopper-Stabilized Stereo Multi-Bit Audio ΣΔ A/D Converter", *ISSCC 2003 Digest of Technical Papers*, Feb. 2003.

21. R.H.M. van Veldhoven, "A Triple-Mode Continuous-Time ΣΔ Modulator with Switched-Capacitor Feedback DAC for a GSM-EDGE/CDMA200/UMTS Receiver", *ISSCC 2003 Digest of Technical Papers*, Feb. 2003.

22. S. Yan and E. Sánchez-Sinencio, "A Continuous-Time ΣΔ Modulator with 88dB Dynamic Range and 1.1MHz Signal Bandwidth", *ISSCC 2003 Digest of Technical Papers*, Feb. 2003.

23. K. Philips, "A 4.4mW 76dB Complex ΣΔ ADC for Bluetooth Receivers", *ISSCC 2003 Digest of Technical Papers*, Feb. 2003.

24. R. Schreier, J. Lloyd, et al., "A 50mW Bandpass ΣΔ ADC with 333kHz BW and 90dB DR", *ISSCC 2002 Digest of Technical Papers*, Feb. 2002.

25. T. Salo, T. Hollman, et al., "A Dual-Mode 80MHz Bandpass ΔΣ Modulator for a GSM/WCDMA IF-Receiver", *ISSCC 2002 Digest of Technical Papers*, Feb. 2002.

26. S.K. Gupta, T.L. Brooks and V. Fong, "A 64MHz ΣΔ ADC with 105dB IM3 Distortion Using a Linearized Replica Sampling Network", *ISSCC 2002 Digest of Technical Papers*, Feb. 2002.

27. T. Burger and Q. Hueng, "A 13.5mW 185 MSamples/s ΔΣ-Modulator for UMTS/GSM Dual-Standard IF Reception", *ISSCC 2001 Digest of Technical Papers*, Feb. 2001.

28. O. Oliaei, P. Clement and P. Gorisse, "A 5mW ΣΔ Modulator with 84dB Dynamic Range for GSM/EDGE", *ISSCC 2001 Digest of Technical Papers*, Feb. 2001.

29. K. Vleugels, S. Rabii and B.A. Wooley, "A 2.5V Broadband Multi-bit ΣΔ Modulator with 95dB Dynamic Range", *ISSCC 2001 Digest of Technical Papers*, Feb. 2001.

30. Y. Geerts, M. Steyaert and W. Sansen, "A 2.5 MSamples/s Multi-Bit ΔΣ CMOS ADC with 95dB SNR", *ISSCC 2000 Digest of Technical Papers*, Feb. 2000.

31. I. Fujimori, L. Longo, et al., "A 90dB SNR, 2.5MHz Output Rate ADC Using Cascaded Multibit ΔΣ Modulation at 8x Oversampling Ratio", *ISSCC 2000 Digest of Technical Papers*, Feb. 2000.

32. A. Tabatabaei, K. Kaviani and B. Wooley, "A Two-Path ΣΔ Modulator with Extended Noise Shaping", *ISSCC 2000 Digest of Technical Papers*, Feb. 2000.
33. Q. Sandifort, L.J. Breems, C. Dijkmans and H. Schuurmans, "IF-to-Digital Converter for FM/AM/IBOC Radio", Proceedings of the 29TH European Solid-State Circuit Conference, Estoril, Portugal, Sept. 2003, pp. 707–710.
34. Catena Radio Design, private communication, 2005.
35. Philips Semiconductors, private communication, 2006.

Harmonic Distortion in CT Integrators Using MOSCAPs

A

The Appendix A presents the calculation of the harmonic distortion at the output of fully-differential CT integrators with MOSFET capacitors (MOSCAPs). The physical properties of MOSCAPs are reviewed in section A.1. The harmonic distortion at the output of a G_m-C integrator with mismatched MOSCAPs is calculated in section A.2. A similar calculation for an integrator with feedback capacitors is shown in section A.3.

A.1 Introduction

High linearity poly-poly capacitors are not available in most standard CMOS processes. In these processes, the only linear capacitors available are poly-metal or metal-metal sandwich structures. In both cases, the capacitance per area is relatively small. The highest possible capacitance per area is achieved by using the thin gate oxide as dielectric, and the MOSFET's gate and bulk (or nwell) as the terminals of a MOSCAP [1]. Figure 5-22 depicts one of the possible MOSCAP structures, the NMOS-on-nwell capacitor used in the fabricated prototypes discussed in Chapter 5. The major drawback of MOSCAPs is the large voltage dependence of the capacitance between the gate and bulk [2].

For an NMOS device, the accumulation region of operation is reached when the device is biased at positive voltages and free electrons accumulate in the channel. Figure A-1 shows how the normalised gate-bulk C_{GB} capacitance varies as function of the voltage V_{GB} applied across the device terminals. This picture also shows the different regions of operation of a MOSFET [2] regarding the channel charge. The following mathematical derivations assume the device is biased in the accumulation region. In this case, after defining $V_{GB} = v + V_B$ (where v is a small-signal voltage and V_B is the dc bias voltage across the device terminals), the MOSCAP capacitance can be expressed as [1]:

$$C(v) = C_{ox} \cdot W \cdot L \cdot \left(1 - \frac{2\phi_t \cdot v}{|V_B - V_{FB}|^2} + \frac{2\phi_t \cdot v^2}{|V_B - V_{FB}|^3} + \dots\right) \quad \text{(A-1)}$$

where C_{ox} is the gate oxide capacitance per unit area, W and L are the device's planar dimensions, ϕ_t is the thermal voltage, and V_{FB} is the flat-band voltage. For the calculations presented in this appendix, Eq. (A-1) is truncated and the notation is simplified to:

$$C(v) \cong c(1 + \alpha_1 \cdot v + \alpha_2 \cdot v^2) \quad \text{(A-2)}$$

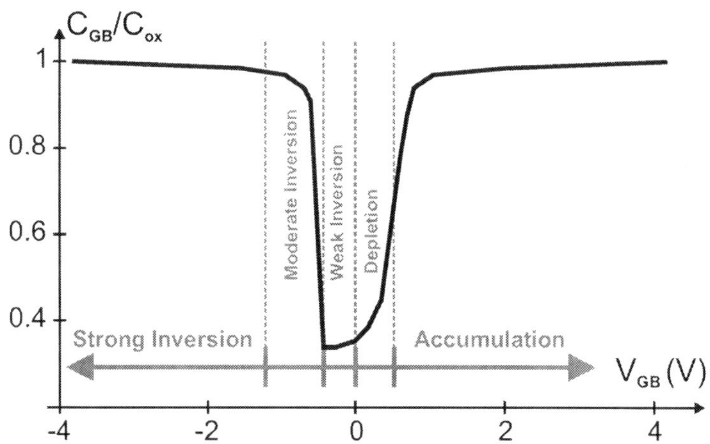

Figure A-1: *Normalised gate-bulk capacitance as a function of the V_{GB} voltage for a NMOS-on-nwell device [1].*

In reference [3], the harmonic distortion due to the V-Q conversion in sampling MOSCAPs is calculated. In SC ΣΔ modulators, only the harmonic distortion taking place at the sampling capacitors of the 1st integrator are relevant. The inter-stage charge transfers are not affected by the MOSCAPs' non-linear capacitance, and the harmonic distortion due to the Q-V conversion taking place at the last stage is strongly attenuated by the noise shaping [3].

In CT ΣΔ modulators with SC DAC, the contribution of the 1st integrating capacitors for the total harmonic distortion is the most relevant among the MOSCAPs used in the implementation. The non-linearity of the feedback DAC capacitors is not important, as they are always charged to the same dc voltages. In the next sections, Eq. (A-2) is used to calculate the harmonic distortion in fully-differential CT integrators due to the Q-V conversion at the integrating capacitors.

A.2 Harmonic Distortion in Differential G_m-C Integrators

Figure A-2 shows a fully-differential G_m-C integrator, implemented with mismatched MOSCAPs. The output CM voltage is positive enough to keep the devices biased in accumulation, and the transconductance G_m is considered perfectly linear in this analysis. As $dq = C(v).dv$, the charge stored in the integrating capacitors can be calculated as a function of the output voltages v_{op} and v_{on} [3]:

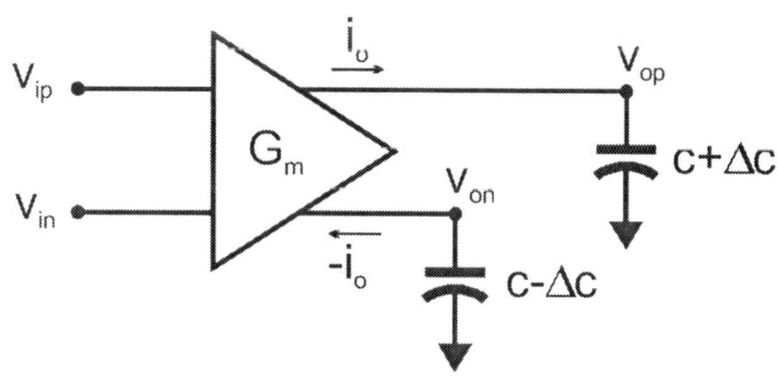

Figure A-2: *Fully differential G_m-C integrator with mismatched MOSCAPs.*

$$q_p(v) = \int_0^{v_{op}} C(v)dv = (c+\Delta c)\cdot\left(v_{op} + \frac{\alpha_1}{2}\cdot v_{op}^2 + \frac{\alpha_2}{3}\cdot v_{op}^3\right) \quad \text{(A-3)}$$

$$q_n(v) = \int_0^{v_{on}} C(v)dv = (c-\Delta c)\cdot\left(v_{on} + \frac{\alpha_1}{2}\cdot v_{on}^2 + \frac{\alpha_2}{3}\cdot v_{on}^3\right) \quad \text{(A-4)}$$

Using the Taylor series approximation, the output voltages v_{op} and v_{on} can be expressed as a function of the stored charges q_p and q_n:

$$v_{op}(q_p) \cong \frac{q_p}{(c+\Delta c)} - \frac{\alpha_1}{2(c+\Delta c)^2}\cdot q_p{}^2 + \frac{3\alpha_1^2 - 2\alpha_2}{6(c+\Delta c)^3}\cdot q_p{}^3 \quad \text{(A-5)}$$

$$v_{on}(q_n) \cong \frac{q_n}{(c-\Delta c)} - \frac{\alpha_1}{2(c-\Delta c)^2}\cdot q_n{}^2 + \frac{3\alpha_1^2 - 2\alpha_2}{6(c-\Delta c)^3}\cdot q_n{}^3 \quad \text{(A-6)}$$

Equations (A-5) and (A-6) reveal that a 2nd-order harmonic distortion term is always present at the output of a single-ended G_m-C integrator implemented with a MOSCAP. As $i(t) = dq/dt$, the small-signal charge variation in the integrating capacitors due to the small-signal output current $i_o(t) = I.\sin(\omega t)$ can be calculated:

$$q(t) = \int_0^{t} I\cdot \sin(\omega\tau)d\tau = \frac{I\cdot(1-\cos(\omega t))}{\omega} \quad \text{(A-7)}$$

The substitution of (A-7) into (A-5) and (A-6) gives output voltages as a function of the output current time-variation:

$$v_{op}(t) \cong \frac{I\cdot(1-\cos(\omega t))}{\omega(c+\Delta c)} - \frac{\alpha_1}{2(c+\Delta c)^2}\cdot\left(\frac{I\cdot(1-\cos(\omega t))}{\omega}\right)^2 \quad \text{(A-8)}$$
$$+ \frac{3\alpha_1^2 - 2\alpha_2}{6(c+\Delta c)^3}\cdot\left(\frac{I\cdot(1-\cos(\omega t))}{\omega}\right)^3$$

$$v_{on}(t) \cong \frac{-I \cdot (1-\cos(\omega t))}{\omega(c-\Delta c)} - \frac{\alpha_1}{2(c-\Delta c)^2} \cdot \left(\frac{I \cdot (1-\cos(\omega t))}{\omega}\right)^2 \qquad \text{(A-9)}$$
$$\frac{3\alpha_1^2 - 2\alpha_2}{6(c-\Delta c)^3} \cdot \left(\frac{I \cdot (1-\cos(\omega t))}{\omega}\right)^3$$

The differential integrator output is defined as:

$$v_o(t) = v_{op}(t) - v_{on}(t) \qquad \text{(A-10)}$$

The substitution of (A-8) and (A-9) into (A-10) results in the following approximation, after some trigonometric manipulation and the elimination of the terms with high-order powers of Δc:

$$v_o(t) \cong (1-\cos(\omega t)) \cdot \left(\frac{2I}{c\omega} + \frac{2I \cdot \Delta c^2}{c^3 \omega} - \frac{2I^2 \Delta c \alpha_1}{c^3 \omega^2}\right) \qquad \text{(A-11)}$$
$$-\cos(2\omega t) \cdot \left(\frac{I^2 \Delta c \alpha_1}{2c^3 \omega^2} - \frac{3I^3 \alpha_1^2}{2c^3 \omega^3} + \frac{I^3 \alpha_2}{c^3 \omega^3}\right) + \cos(3\omega t) \cdot \left(\frac{I^3 \alpha_2}{6c^3 \omega^3} - \frac{I^3 \alpha_1^2}{4c^3 \omega^3}\right)$$

In Eq. (A-11), the approximation $(c+\Delta c)(c-\Delta c) \approx c^2$ is also used to simplify the denominator of the resulting expression. The 2nd-order (HD_2) and 3rd-order (HD_3) harmonic distortion coefficients at the integrator output are then given by:

$$HD_2 \cong \frac{I \cdot \Delta c \cdot \alpha_1}{4 \cdot c^2 \cdot \omega} \qquad \text{(A-12)}$$

$$HD_3 \cong \frac{I^2 (3 \cdot \alpha_1^2 - 2 \cdot \alpha_2)}{24 \cdot c^2 \cdot \omega^2} \qquad \text{(A-13)}$$

Because of the fully-differential structure, if there is no mismatch between the integrating MOSCAPs, the 2nd-order harmonic distortion term due to the Q-V conversion is completely cancelled.

A.3 Harmonic Distortion in Differential Feedback Integrators

Figure A-3 shows a fully-differential integrator with feedback capacitors. The floating capacitors are implemented with series-connected MOSCAPs, as depicted in Figure 5.23. The CM mode voltages at the input virtual-ground nodes and at the integrator outputs are, respectively, V_{CM1} and V_{CM2}. The bias voltage V_B guarantees a positive voltage difference from gate to bulk in all MOSCAPs. In this example, the capacitance of the four integrating MOSCAPs are defined as:

$$C_{1A}(v) = C_{2B}(v) \cong (c + \Delta c)(1 + \alpha_1 \cdot v + \alpha_2 \cdot v^2) \qquad \text{(A-14)}$$
$$C_{1B}(v) = C_{2A}(v) \cong (c - \Delta c)(1 + \alpha_1 \cdot v + \alpha_2 \cdot v^2)$$

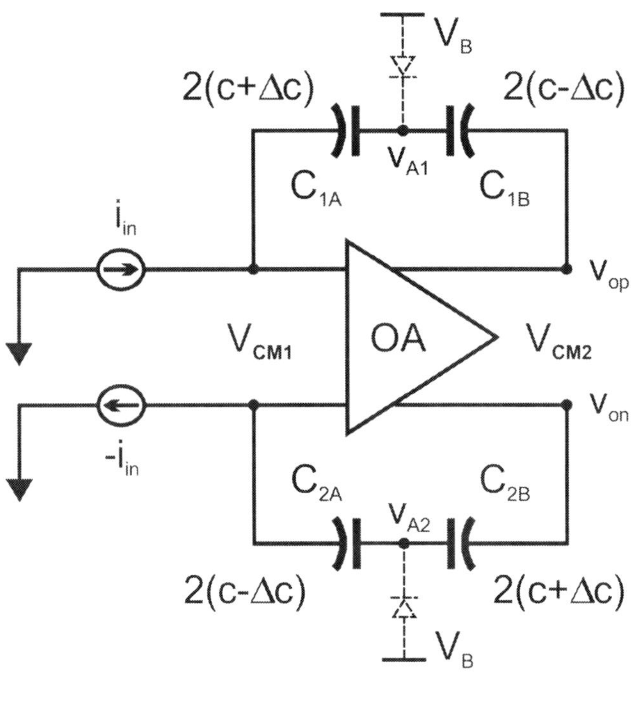

Figure A-3: *Fully-differential CT feedback integrator with series-connected mismatched MOSCAPs.*

Due to the charge conservation is nodes A1 and A2, the total charges $q_p(v)$ and $q_n(v)$, respectively stored in the equivalent series capacitors C_1 and C_2, can be expressed as [3]:

$$q_p(v) = \frac{1}{2}\left[\int_{(V_B - V_{CM1})}^{(v_{A1} - V_{CM1})} 2C_{1A}(v)dv + \int_{(V_B - V_{CM2})}^{(v_{A1} - v_{op})} 2C_{1B}(v)dv\right] \quad \text{(A-15)}$$

$$q_n(v) = \frac{1}{2}\left[\int_{(V_B - V_{CM1})}^{(v_{A2} - V_{CM1})} 2C_{2A}(v)dv + \int_{(V_B - V_{CM2})}^{(v_{A2} - v_{on})} 2C_{2B}(v)dv\right] \quad \text{(A-16)}$$

After defining $V_{CM1}=\Delta V$ and $V_{CM2}=0$, the calculation of the integrals in (A-15) and (A-16) results in the following expressions for $q_p(v)$ and $q_n(v)$:

$$q_p(v) = (\Delta c - c)\left[v_{A1} - V_B - v_{op} + \frac{\alpha_1[(v_{A1}-v_{op})^2 - V_B^2]}{2} + \frac{\alpha_2[(v_{A1}-v_{op})^3 - V_B^3]}{3}\right] \quad \text{(A-17)}$$
$$+(c+\Delta c)\left[v_{A1} - V_B + \frac{\alpha_1[(v_{A1}-\Delta V)^2 - (V_B - \Delta V)^2]}{2} + \frac{\alpha_2[(v_{A1}-\Delta V)^3 - (V_B-\Delta V)^3]}{3}\right]$$

$$q_n(v) = -(c+\Delta c)\left[v_{A2} - V_B - v_{on} + \frac{\alpha_1[(v_{A2}-v_{on})^2 - V_B^2]}{2} + \frac{\alpha_2[(v_{A2}-v_{on})^3 - V_B^3]}{3}\right] \quad \text{(A-18)}$$
$$+(c-\Delta c)\left[v_{A2} - V_B + \frac{\alpha_1[(v_{A2}-\Delta V)^2 - (V_B - \Delta V)^2]}{2} + \frac{\alpha_2[(v_{A2}-\Delta V)^3 - (V_B-\Delta V)^3]}{3}\right]$$

Closed expressions for $q_p(v)$ and $q_n(v)$, that are just functions of the integrator's output voltages, can be obtained if the small-signal voltages v_{A1} and v_{A2} at nodes A1 and A2 are approximated as [3]:

$$v_{A1} = V_B + \frac{v_{op}}{2}, \quad v_{A2} = V_B + \frac{v_{on}}{2} \quad \text{(A-19)}$$

The substitution of (A-19) in (A-17) and (A-18) results in:

$$q_p(v_{op}) = v_{op}(k_1 c + k_2 \Delta c) + \frac{v_{op}^2}{4}(k_3 \Delta c - c \Delta V \alpha_2) + \frac{v_{op}^3 \cdot c\alpha_2}{12} \quad \text{(A-20)}$$

$$q_n(v_{on}) = v_{on}(k_1 c - k_2 \Delta c) - \frac{v_{on}^2}{4}(k_3 \Delta c + c \Delta V \alpha_2) + \frac{v_{on}^3 \cdot c\alpha_2}{12} \quad \text{(A-21)}$$

where the time-invariant terms in (A-17) and (A-18) have been rearranged as the constants k_1, k_2 and k_3:

$$k_1 = 1 + V_B \alpha_1 + V_B^2 \alpha_2 - V_B \Delta V \alpha_2 + \frac{\Delta V^2 \alpha_2 - \Delta V \alpha_1}{2} \quad \text{(A-22)}$$

$$k_2 = \frac{(\Delta V^2 \alpha_2 - \Delta V \alpha_1)}{2} - V_B \Delta V \alpha_2 \quad \text{(A-23)}$$

$$k_3 = \alpha_1 - \Delta V \alpha_2 + 2 V_B \alpha_2 \quad \text{(A-24)}$$

Once again, the use of Taylor series approximations on (A-20) and (A-21) allows the output voltages v_{op} and v_{on} to be written as a function of the stored charges q_p and q_n:

$$v_{op} \cong \frac{q_p}{ck_1 + \Delta c k_2} + q_p^2 \cdot \frac{c \Delta V \alpha_2 - 4 k_3 \Delta c}{4(ck_1 + \Delta c k_2)^3} \quad \text{(A-25)}$$
$$+ q_p^3 \cdot \frac{48 k_3^2 \Delta c^2 - 2c^2 k_1 \alpha_2 + 3c^2 \Delta V^2 \alpha_2^2 - 2ck_2 \alpha_2 \Delta c - 24 c k_3 \Delta V \alpha_2 \Delta c}{24(ck_1 + \Delta c k_2)^5}$$

$$v_{on} \cong \frac{q_n}{ck_1 - \Delta c k_2} + q_n^2 \cdot \frac{c \Delta V \alpha_2 + 4 k_3 \Delta c}{4(ck_1 - \Delta c k_2)^3} \quad \text{(A-26)}$$
$$+ q_n^3 \cdot \frac{48 k_3^2 \Delta c^2 - 2c^2 k_1 \alpha_2 + 3c^2 \Delta V^2 \alpha_2^2 + 2ck_2 \alpha_2 \Delta c + 24 c k_3 \Delta V \alpha_2 \Delta c}{24(ck_1 - \Delta c k_2)^5}$$

Different from single-ended G_m-C integrators using a grounded MOSCAP, the 2nd-order harmonic distortion term at the output of a single-ended feedback integrator implemented with series-connected MOSCAPs is cancelled if $\Delta V = \Delta c = 0$.

The differential integrator output is defined again as:

$$v_o(t) = v_{op}(t) - v_{on}(t) \tag{A-27}$$

The substitution of (A-7), (A-25) and (A-26) in (A-27), following some trigonometric manipulation and the elimination of the terms with high-order powers of Δc, results in a simplified expression for the differential integrator output:

$$v_o(t) \cong \frac{2I}{c\omega k_1} - \frac{3I^2 \Delta c k_3}{c^3 \omega^2 k_1^3} - \cos(\omega t) \cdot \left(\frac{2I}{c\omega k_1} + \frac{4I^2 \Delta c k_3}{c^3 \omega^2 k_1^3} - \frac{3I^2 \Delta c \Delta V k_2 \alpha_2}{c^3 \omega^2 k_1^4} \right) \tag{A-28}$$
$$- \cos(2\omega t) \cdot \left(\frac{I^2 \Delta c k_3}{c^3 \omega^2 k_1^3} - \frac{3I^2 \Delta c \Delta V k_2 \alpha_2}{4c^3 \omega^2 k_1^4} \right) + \cos(3\omega t) \cdot \left(\frac{I^3 \Delta V^2 \alpha_2^2}{16c^3 \omega^3 k_1^5} - \frac{I^3 \alpha_1^2}{24c^3 \omega^3 k_1^4} \right)$$

Based on (A-28), the 2nd-order and 3rd-order harmonic distortion coefficients at the integrator output are calculated:

$$HD_2 \cong \frac{I \cdot \Delta c \cdot (4k_1 k_3 + 3k_2 \Delta V \alpha_2)}{8 \cdot c^2 \cdot \omega \cdot k_1^3} \tag{A-29}$$

$$HD_3 \cong \frac{I^2(3 \cdot \Delta V^2 \cdot \alpha_2^2 - 2 \cdot k_1 \cdot \alpha_2)}{96 \cdot c^2 \cdot \omega^2 \cdot k_1^4} \tag{A-30}$$

The 2nd-order harmonic distortion coefficient is a function of the relative mismatch Δc among the MOSCAPs, and of the voltage difference ΔV between the input and output CM voltages. The dominant terms for the determination of the 3rd-order harmonic distortion coefficient do not depend on Δc. In the ΣΔ modulator design presented in Chapter 5, ΔV = 0.8V - 0.7V = 0.1V and Δc depends on the actual lay-out of the MOSCAPs used in the 1st loop filter integrator.

References

1. A.T. Behr, M.C. Schneider, S. Noceti Filho and C.G. Montoro, "Harmonic Distortion Caused by Capacitors Implemented with MOSFET Gates", *IEEE J. Solid-State Circ.*, vol. 27, 1470–1475, Oct. 1992.

2. Y. Tsividis, "Operation and Modeling of the MOS Transistor", 2nd edition, Oxford University Press, Oxford, 1999.
3. H. Yoshizawa, Y. Huang, P.F. Ferguson Jr. and G.C. Temes, "MOSFET-Only Switched-Capacitor Circuits in Digital CMOS Technology", *IEEE J. Solid-State Circ.*, vol. 34, 734–747, June 1999.

Noise Analysis of CT ΣΔ Modulators with SC Feedback DAC

B

The Appendix B presents the calculation of the input-referred noise in CT ΣΔ modulators with SC feedback DAC. Section B.1 reviews the input-referred noise calculation in standard 1st-order CT ΣΔ modulator with CT feedback. This result can be generalised to high-order modulators as the dominant thermal noise source is often the front-end circuitry. The calculation of the noise power spectrum density (PSD) across a switched capacitor is reviewed in section B.2. The noise contribution of the SC MOS switches (and other resistive noise sources) is calculated in section B.3. The noise contributions from the SC DAC reference voltage and from the opamp are calculated, respectively, in sections B.4 and B.5. The final conclusions are drawn in section B.6

B.1 Introduction

CT modulators have important advantages over their SC counterparts: they do not require a sample-and-hold front-end, as the loop filter acts as an anti-aliasing filter, and can be operated at higher sampling frequencies (f_S) [1]. However, CT modulators are more sensitive to clock jitter than SC implementations. To overcome this limitation, the combination of a CT loop filter with a SC feedback DAC has been proposed [2].

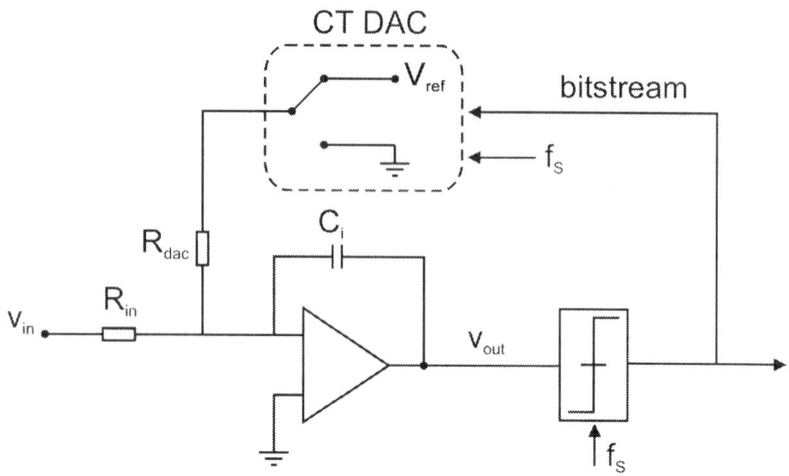

Figure B-1: *1st-order CT ΣΔ modulator with CT feedback.*

Accounting for the dominant noise sources in the circuit implementation of a ΣΔ modulator is essential for a successful design. The dominant noise sources are determined by the loop filter topology and the scaling of the coefficients. Usually, they originate from the circuits at the modulator's front-end: the 1st integrator and the feedback DAC. In this case, the calculations based on 1st-order modulators are also valid for high-order modulators.

Figure B-1 shows the schematic diagram of a 1st-order single-bit CT modulator with a CT feedback DAC. The dominant noise sources are the input resistor R_{in}, the DAC resistor R_{dac}, the DAC reference voltage V_{ref}, and the opamp. The input-referred circuit noise PSD of the ΣΔ modulator is given by:

$$S_{in}(f) = 4k\theta R_{in} \cdot \left(1 + \frac{R_{in}}{R_{dac}}\right) + S_{Vref} \cdot \left(\frac{R_{in}}{R_{dac}}\right)^2 + S_{OA} \cdot \left[(\omega \cdot R_{in} C_i)^2 + \left(\frac{R_{dac} + R_{in}}{R_{dac}}\right)^2\right] \quad \text{(B-1)}$$

where k is the Boltzman's constant, θ is the absolute temperature in Kelvin, $S_{Vref}(f)$ is the reference voltage output noise PSD, and $S_{OA}(f)$ is the equivalent opamp input noise PSD. For the sake of simplicity, the opamp' s *1/f* noise is not accounted in (B-1).

B.2 Noise Voltage PSD Across a Switched Capacitor

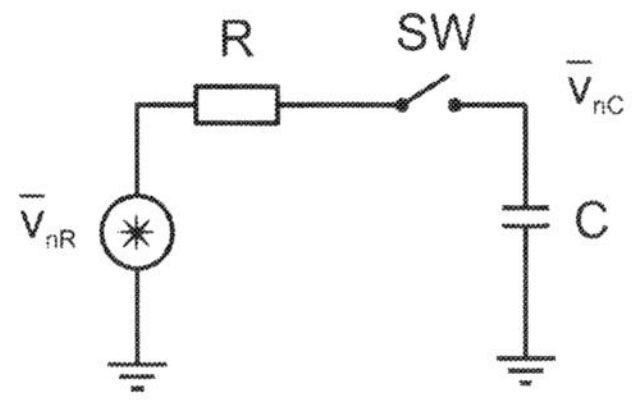

Figure B-2: *Equivalent circuit to evaluate the noise PSD across a sampling capacitor C due to a resistance R.*

The PSD of the noise voltage across a switched capacitor can be calculated from the equivalent circuit presented in Figure B-2. The circuit consists of a MOS switch and an ideal capacitor *C*. The MOS switch is modelled as a combination of an ideal switch *SW*, a series resistance *R* and its associated thermal noise voltage ($\overline{v}_{nR}$). The resistor' s thermal noise PSD (single-sided) is given by:

$$S_R(f) = 4k\theta R \tag{B-2}$$

Assuming that the switch opens/closes with a periodicity T_S and duty cycle τ, the noise voltage PSD across the switched-capacitor present two uncorrelated components [3]. The first component is due to the noise source when it is directly coupled to the capacitor, called "direct noise". The second component, due to the noise sampled-and-held (S/H) in the capacitor when the switch is open, is called "S/H noise". The "direct noise" PSD $S_{RC}(f)$ is the noise obtained after lowpass filtering the noise source $\overline{v}_{nR}$ with the RC network formed by the switch resistance and the capacitor:

$$S_{RC}(f) = \tau \cdot \frac{4k\theta R}{1 + (2\pi f R C)^2} \tag{B-3}$$

In general, the cut-off frequency of the band limiting RC network (f_C) is much higher than f_S. In this case, the "S/H noise" component PSD $S_{SH}(f)$ is determined by the undersampling and aliasing of the resistor' s 1st-order filtered thermal noise PSD $S_{RC}(f)$.

It is demonstrated in [3] that, for any ratio f_C/f_S, $S_{SH}(f)$ can be expressed as:

$$S_{SH}(f) = \frac{\pi f_C}{f_S} \cdot S_R \cdot \frac{\sinh\left(\frac{2\pi f_C}{f_S}\right)}{\cosh\left(\frac{2\pi f_C}{f_S}\right) - \cos\left(\frac{2\pi f}{f_S}\right)} \cdot (1-\tau)^2 \cdot \mathrm{sinc}^2\left(\frac{\pi(1-\tau)f}{f_S}\right) \quad \text{(B-4)}$$

For $f_C > f_S$ and f_C given by $1/2\pi RC$, (B-4) represents an white noise PSD multiplied by a $\mathrm{sinc}^2()$ factor and can be rewritten as:

$$S_{SH}(f) = \frac{(1-\tau)^2}{f_S} \cdot \frac{2k\theta}{C} \cdot \mathrm{sinc}^2\left(\frac{\pi(1-\tau)f}{f_S}\right) \quad \text{(B-5)}$$

Equations (B-3) and (B-5) are the basic expressions required to calculate the output noise PSD in any kind of SC network and they are used throughout the next sections.

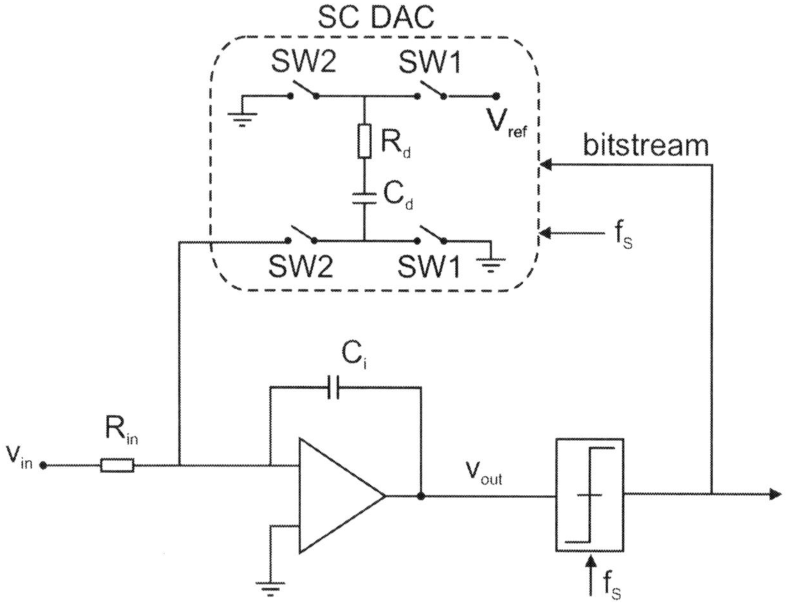

Figure B-3: *1st-order CT ΣΔ modulator with SC feedback.*

B.3 Input-Referred Thermal Noise Due To the SC Feedback DAC MOS Switches

Figure B-3 shows a CT ΣΔ modulator with a SC feedback DAC. The timing diagram of the SC DAC is presented in Figure B-4. During phase I, the switches *SW1* are closed and the DAC capacitor C_d is charged. In this single-ended model, C_d is connected to the input virtual-ground node of the ΣΔ modulator during phase III, if bitstream is 1. During phases II and IV, all switches are open.

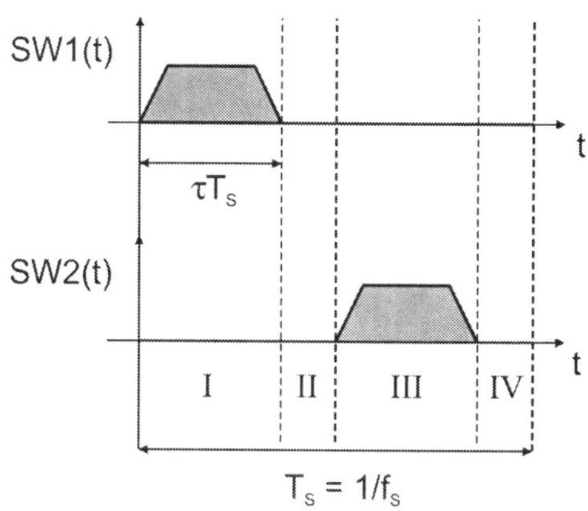

Figure B-4: *Timing diagram for the SC DAC.*

Based on the methodology introduced by [4], the "direct noise" and "S/H noise" components due to resistive noise sources are calculated in the following paragraphs. After the evaluation of the total noise PSD at the output of the ΣΔ modulator's 1st integrator, the input-referred noise is determined. The DAC reference voltage and the opamp are considered noiseless for the calculations in this section.

B.3.1 "Direct Noise" Components

The "direct noise" components at the output of the 1st integrator are calculated based on the equivalent noise circuit of the front-end during the different clocking phases of the SC DAC. During phases I, II and IV the SC network is disconnected from the integrator circuit (Figure B-5a). The output noise PSD is then:

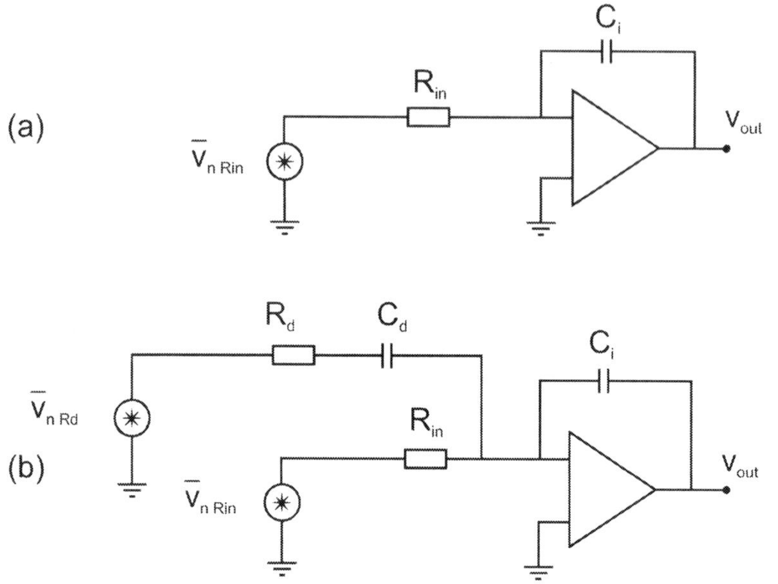

Figure B-5: *Equivalent circuit to calculate the "direct noise" contribution during phases I, II and IV (a), and during phase III (b).*

$$S_{D1}(f) = (1-\tau)\cdot\frac{4k\theta R_{in}}{(2\pi f R_{in} C_i)^2} \qquad \text{(B-6)}$$

The equivalent noise circuit during phase III is presented in Figure B-5b. In this case the total noise PSD at output of the CT integrator is given by:

$$S_{D2}(f) = \tau\cdot\left[\frac{4k\theta R_{in}}{(2\pi f R_{in} C_i)^2} + \frac{G^2\cdot 4k\theta R_d}{1+(2\pi f R_d C_d)^2}\right] \qquad \text{(B-7)}$$

where $G = C_d/C_i$ is the ratio between the sampling and the integrating capacitors, and R_d is the total series resistance of the discharging path, including the on-resistances (r_{on}) of the two MOS switches.

B.3.2 'Sampled-and-Held" Components

The thermal noise voltage from the series resistance R_d is sampled-and-held in C_d and C_i, and then made available at the output of the integrator. Figure B-6a shows the equivalent circuit used to calculate the "S/H noise" contribution from switches *SW1*. The noise PSD across C_d in phase II, after switches *SW1* are open, is given by (B-5). The sampled noise is then integrated by a SC integrator comprised by the SC network, C_i and the opamp. The squared modulus of the transfer function of a discrete-time Euler integrator is [4]:

$$|H_{SCint}(f)|^2 = \frac{G^2}{4 \cdot \sin(\pi f/ f_S)^2} \quad \text{(B-8)}$$

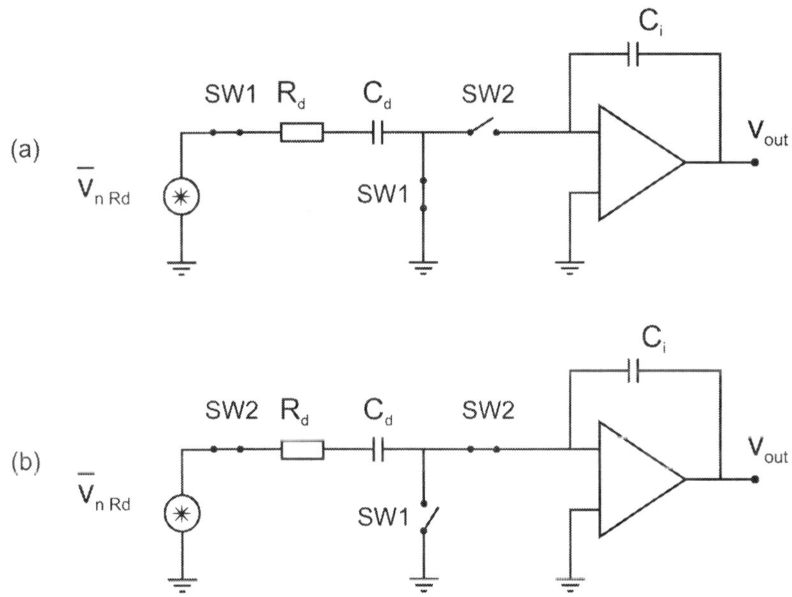

Figure B-6: *Equivalent circuit to calculate "S/H noise" contribution from SW1 switches (a), and from SW2 switches (b).*

The combination of (B-5) and (B-8) gives the output noise PSD due to the discrete-time integrated "S/H noise" contribution from switches *SW1*. Because the integrator output is held with almost 100% duty cycle, then it is possible to approximate τ=0:

$$S_{SH1}(f) = \frac{G^2}{4 \cdot \sin(\pi f / f_S)^2} \cdot \frac{2k\theta}{f_S \cdot C_d} \cdot \operatorname{sinc}\left(\frac{\pi f}{f_S}\right)^2 \quad \text{(B-9)}$$

Figure B-6b shows the equivalent circuit to calculate the sampled-and-held noise contribution from switches *SW2*. Two "S/H noise" contributions are present at the output of the integrator. The first PSD $S_{S2}(f)$ is the noise voltage sampled-and-held in C_i during phase IV, after the switches *SW2* are open:

$$S_{SH2}(f) = G^2 \cdot \frac{(1-\tau)^2}{f_S} \cdot \frac{2k\theta}{C_d} \cdot \operatorname{sinc}\left(\frac{\pi(1-\tau)f}{f_S}\right)^2 \quad \text{(B-10)}$$

The second contribution $S_{S3}(f)$ is the noise voltage sampled-and-held in C_d after the switches *SW2* are open, and discrete-time integrated in C_i during the next clock cycle:

$$S_{SH3}(f) = \frac{G^2}{4 \cdot \sin(\pi f / f_S)^2} \cdot \frac{2k\theta}{f_S \cdot C_d} \cdot \operatorname{sinc}\left(\frac{\pi f}{f_S}\right)^2 \quad \text{(B-11)}$$

B.3.3 Input-Referred Noise

To calculate the input-referred noise PSD, it is necessary to determine the total output noise PSD and divide it by the squared modulus transfer function of the CT integrator. The latter is given by:

$$|H_{CTint}(f)|^2 = \frac{1}{(2\pi f R_{in} C_i)^2} \quad \text{(B-12)}$$

In order to simplify our results, we assume a high oversampling ratio (*OSR*). That assumption is valid for most ΣΔ modulator applications. In this case, the following approximations hold:

$$\sin(x) \cong x, \qquad \operatorname{sinc}(x) \cong 1 \quad \text{(B-13)}$$

Other simplifications are possible in the frequencies where the 1st integrator gain is high. In (B-7), the "direct noise" contribution of R_d is negligible compared to the noise contribution of R_{in}. The "S/H noise" component (B-10) is negligible compared to (B-9) and (B-11). The total "direct noise" and "S/H noise" components at the output of the integrator can be, respectively, approximated by:

$$S_D(f) \cong \frac{4k\theta R_{in}}{(2\pi f R_{in} C_i)^2} \qquad \text{(B-14)}$$

$$S_{SHT}(f) \cong \frac{G^2 \cdot f_S^2}{(2\pi f)^2} \cdot \frac{4k\theta}{f_S \cdot C_d} \qquad \text{(B-15)}$$

The division of (B-14) and (B-15) by (B-12) gives the input-referred noise PSD of the hybrid CT/SC ΣΔ modulator due to the resistive noise sources:

$$S_{inSW}(f) \cong 4k\theta R_{in} + 4k\theta R_{in}^2 \cdot \frac{C_d}{T_S} \qquad \text{(B-16)}$$

It is easy to recognize the (T_S/C_d) factor as the expression of the equivalent resistor of a switched-capacitor network:

$$R_{dac} = T_S/C_d \qquad \text{(B-17)}$$

Combining (B-16) and (B-17) we obtain:

$$S_{inSW}(f) = 4k\theta R_{in} \cdot \left[1 + \frac{R_{in}}{R_{dac}}\right] \qquad \text{(B-18)}$$

B.4 Input-Referred Thermal Noise Due To the SC Feedback DAC Reference Voltage

The output noise from the DAC reference voltage does not have a direct path to the output, and just a sampled-and-held noise component is integrated. The equivalent circuit for the S/H noise calculation is show in Figure B-7. Because the output noise of the reference voltage is normally lowpass filtered by a large external capacitor, $S_{Vref}(f)$ is not undersampled.

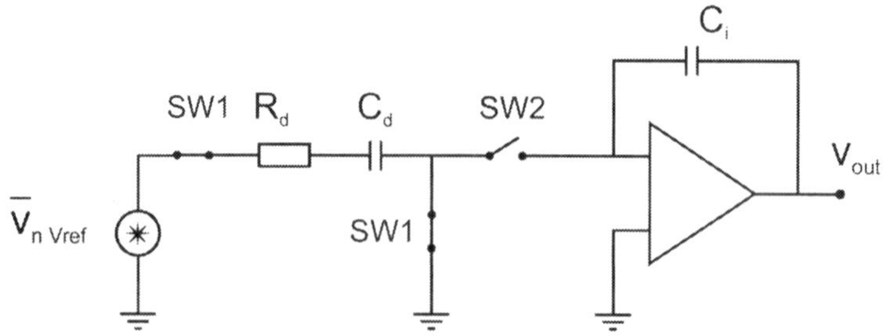

Figure B-7: *Equivalent circuit to calculate "Sampled Noise" contribution from the reference voltage.*

In this case, the DT integrated output S/H noise PSD is simply:

$$S_{S4}(f) = \frac{G^2}{4 \cdot \sin(\omega/2f_S)^2} \cdot \sum_{m=-\infty}^{\infty} S_{Vref}(f - mf_S) \cdot \mathrm{sinc}^2\left(\frac{\omega}{2f_S}\right) \tag{B-19}$$

The input-referred noise is then calculated in the same way as in the previous section:

$$S_{inVref}(f) \cong (R_{in} \cdot C_d \cdot f_S)^2 \cdot S_{Vref} = \left(\frac{R_{in}}{R_{dac}}\right)^2 \cdot S_{Vref} \tag{B-20}$$

B.5 Input-Referred Thermal Noise Due To the OpAmp

The equivalent circuit for the calculation of the opamp's direct noise contribution is shown in Figure B-8a. During phases I, II and IV, just R_{in} is connected to the virtual-ground node. During phase III, the R_d-C_d branch is connected in parallel to R_{in}. As the impedance of the R_d-C_d series combination is very high for $\omega < 1/(R_d+R_{in})C_d$, the low-frequency opamp's direct noise contribution at the integrator's output during all clock phases can be approximated by:

$$S_{D3} = \frac{\omega^2 R_{in}^2 C_i^2 + 1}{\omega^2 R_{in}^2 C_i^2} \cdot S_{OA} \tag{B-21}$$

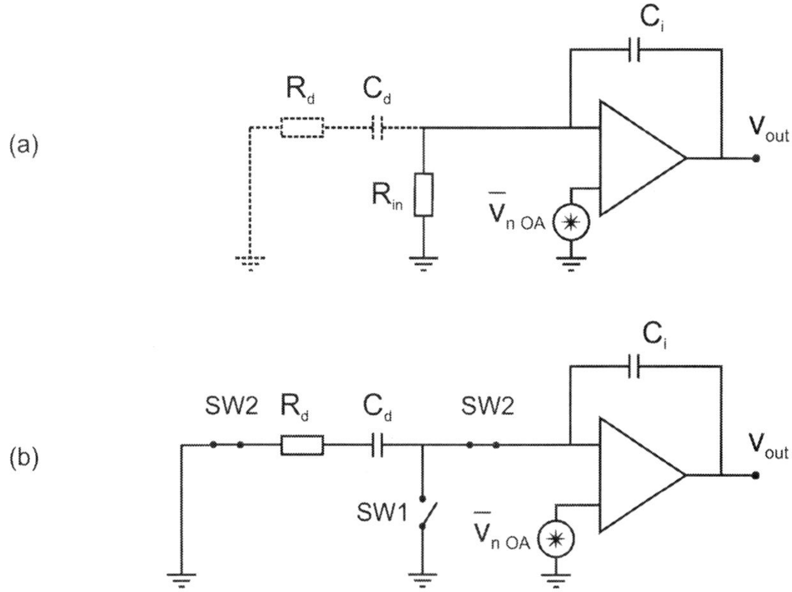

Figure B-8: *Equivalent circuit to calculate the opamp's "direct noise" contribution during all phases (a), and "S/H noise" in capacitors C_d and C_i (b).*

Assuming an opamp with an unity gain frequency much larger than f_C, the transfer function from the opamp input to the sampling capacitor C_d can be approximated by a 1st-order filter:

$$\frac{v_{Cd}}{v_{nOA}} = \frac{1}{sR_dC_d + 1} \tag{B-22}$$

The noise PSD at the output due to the opamp noise sampled-and-held in C_d and discrete-time integrated is then given by:

$$S_{S5} = \frac{G^2}{4 \cdot \sin(\omega / 2f_S)^2} \cdot \frac{\pi S_{OA}}{f_S} \cdot \frac{1}{2\pi R_d C_d} \cdot \operatorname{sinc}\left(\frac{\omega}{2f_S}\right)^2 \tag{B-23}$$

The noise PSD sampled-and-held in C_i is not discrete-time integrated and therefore is not dominant for the input-referred noise calculation.

The input-referred noise due to the opamp can be approximated at low frequencies by:

$$S_{inOA}(f) \cong \left(\omega^2 R_{in}^2 C_i^2 + 1 + \frac{C_d \cdot f_S \cdot R_{in}^2}{2R_d}\right) \cdot S_{OA} = \left(\omega^2 R_{in}^2 C_i^2 + \frac{2R_d R_{dac} + R_{in}^2}{2R_d R_{dac}}\right) \cdot S_{OA}$$

(B-24)

B.6 Conclusion

The total input-referred noise PSD of the CT ΣΔ modulator with SC feedback DAC, under a high *OSR* condition, is determined after grouping Eqs. (B-18), (B-20) and (B-24):

$$S_{in}(f) = 4k\theta R_{in} \cdot \left(1 + \frac{R_{in}}{R_{dac}}\right) + S_{Vref} \cdot \left(\frac{R_{in}}{R_{dac}}\right)^2 + S_{OA} \cdot \left[(\omega \cdot R_{in} C_i)^2 + \frac{2R_d R_{dac} + R_{in}^2}{2R_d R_{dac}}\right]$$

(B-25)

Equation (B-25) is very similar to (B-1), with the exception of the opamp low-frequency noise contribution. For the parameters in this design (R_{in}=1,650 kΩ, R_{dac}=5400 kΩ and R_d=265 Ω), the low-frequency opamp noise contribution is only 15% higher in the SC feedback case.

References

1. E.J. van der Zwan and E.C. Dijkmans, "A 0.2-mW CMOS ΣΔ Modulator for Speech Coding with 80dB Dynamic Range", *IEEE J. Solid-State Circ.*, vol. 31, 1873–1880, Dec. 1996.
2. M. Ortmanns, F. Gerfers and Y. Manoli, "A Continuous-Time ΣΔ Modulator with Reduced Sensitivity to Clock Jitter Through SCR Feedback", *IEEE Trans. CAS-I.*, vol. 52, 875–884, May 2005.
3. C. Gobet and A. Knob, "Spectral Distribution of a Sampled 1st-order Lowpass Filtered White Noise", *Electron. Lett.*, vol. 17, Sept. 1981.
4. C. Gobet and A. Knob, "Noise Analysis of Switched Capacitor Networks", *IEEE Trans. Circ. Syst.*, vol. 30, 37–43, Jan. 1983.

List of Acronyms

ABS	automatic braking system
ac	alternate current
A/D	analog-to-digital
ADC	analog-to-digital converter
AGC	automatic gain control
AM	amplitude modulation
BW	bandwidth
CB	citizens band
CLK	clock
CM	cross-modulation
CMOS	complementary metal-oxide semiconductor
CPC	complementary pair convolution
CT	continuous-time
D/A	digital-to-analog
DAB	digital audio broadcasting
DAC	digital-to-analog converter
dc	direct current
DEM	dynamic element matching
DLL	delay-locked loop
DR	dynamic range
DSP	digital signal processor
DT	discrete-time
ENOB	effective number of bits
FEC	forward error correction
FFT	fast fourier transform
FCC	federal communications commission
FM	frequency modulation
G_m-C	transconductance-capacitor
HD	harmonic distortion

f_{RF}, ω_{RF}	input RF frequency
f_S, ω_S	sampling frequency
f_u, ω_u	OpAmp unity-gain frequency
$f_S/2$	half sampling frequency
g_m	transistor small-signal transconductance
G_m	OTA transconductance
I_b	bias current
I_D	drain current
I_{DAC}	DAC current
k	Boltzman's constant
L	channel length of MOSFET
$L(s)$	loop filter transfer function
n^2_{rms}	in-band shaped quantization noise power
N	loop filter order
$N_Q(f)$	shaped quantization noise psd
q^2_{rms}	total quantization noise power
$Q(f)$	quantization noise psd
r_{on}	MOSFET switch on-resistance
R_d	SC DAC series resistor
R_{dac}	CT DAC resistor
R_{dg}	degeneration resistor
R_{in}	input resistor
R_s	signal source output resistance
T_S	sampling period
V_{dac}	DAC feedback voltage
V_{dc}	input offset voltage
VDD	positive supply voltage of MOS circuits
V_{DS}	drain-source voltage
V_g	residual voltage across the OTA inputs
V_{GS}	gate-source voltage
V_{GT}	gate overdrive voltage
V_{ref}	dac reference voltage
VSS	negative supply voltage of MOS circuits
V_T	threshold voltage
W	channel width of a MOSFET

Index

A

B

C

D

E

Printed in the United States
106171LV00002B/163-177/P

9 781402 081637